The Gun in Central Africa

NEW AFRICAN HISTORIES

SERIES EDITORS: JEAN ALLMAN, ALLEN ISAACMAN, AND DEREK R. PETERSON

Books in this series are published with support from the
Ohio University Center for International Studies.

David William Cohen and E. S. Atieno Odhiambo, *The Risks of Knowledge: Investigations into the Death of the Hon. Minister John Robert Ouko in Kenya, 1990*

Belinda Bozzoli, *Theatres of Struggle and the End of Apartheid*

Gary Kynoch, *We Are Fighting the World: A History of the Marashea Gangs in South Africa, 1947–1999*

Stephanie Newell, *The Forger's Tale: The Search for Odeziaku*

Jacob A. Tropp, *Natures of Colonial Change: Environmental Relations in the Making of the Transkei*

Jan Bender Shetler, *Imagining Serengeti: A History of Landscape Memory in Tanzania from Earliest Times to the Present*

Cheikh Anta Babou, *Fighting the Greater Jihad: Amadu Bamba and the Founding of the Muridiyya in Senegal, 1853–1913*

Marc Epprecht, *Heterosexual Africa? The History of an Idea from the Age of Exploration to the Age of AIDS*

Marissa J. Moorman, *Intonations: A Social History of Music and Nation in Luanda, Angola, from 1945 to Recent Times*

Karen E. Flint, *Healing Traditions: African Medicine, Cultural Exchange, and Competition in South Africa, 1820–1948*

Derek R. Peterson and Giacomo Macola, editors, *Recasting the Past: History Writing and Political Work in Modern Africa*

Moses E. Ochonu, *Colonial Meltdown: Northern Nigeria in the Great Depression*

Emily S. Burrill, Richard L. Roberts, and Elizabeth Thornberry, editors, *Domestic Violence and the Law in Colonial and Postcolonial Africa*

Daniel R. Magaziner, *The Law and the Prophets: Black Consciousness in South Africa, 1968–1977*

Emily Lynn Osborn, *Our New Husbands Are Here: Households, Gender, and Politics in a West African State from the Slave Trade to Colonial Rule*

Robert Trent Vinson, *The Americans Are Coming! Dreams of African American Liberation in Segregationist South Africa*

James R. Brennan, *Taifa: Making Nation and Race in Urban Tanzania*

Benjamin N. Lawrance and Richard L. Roberts, editors, *Trafficking in Slavery's Wake: Law and the Experience of Women and Children*

David M. Gordon, *Invisible Agents: Spirits in a Central African History*

Allen F. Isaacman and Barbara S. Isaacman, *Dams, Displacement, and the Delusion of Development: Cahora Bassa and Its Legacies in Mozambique, 1965–2007*

Stephanie Newell, *The Power to Name: A History of Anonymity in Colonial West Africa*

Gibril R. Cole, *The Krio of West Africa: Islam, Culture, Creolization, and Colonialism in the Nineteenth Century*

Matthew M. Heaton, *Black Skin, White Coats: Nigerian Psychiatrists, Decolonization, and the Globalization of Psychiatry*

Meredith Terretta, *Nation of Outlaws, State of Violence: Nationalism, Grassfields Tradition, and State Building in Cameroon*

Paolo Israel, *In Step with the Times: Mapiko Masquerades of Mozambique*

Michelle R. Moyd, *Violent Intermediaries: African Soldiers, Conquest, and Everyday Colonialism in German East Africa*

Abosede A. George, *Making Modern Girls: A History of Girlhood, Labor, and Social Development in Colonial Lagos*

Alicia C. Decker, *In Idi Amin's Shadow: Women, Gender, and Militarism in Uganda*

Rachel Jean-Baptiste, *Conjugal Rights: Marriage, Sexuality, and Urban Life in Colonial Libreville, Gabon*

Shobana Shankar, *Who Shall Enter Paradise? Christian Origins in Muslim Northern Nigeria, ca. 1890–1975*

Emily S. Burrill, *States of Marriage: Gender, Justice, and Rights in Colonial Mali*

Todd Cleveland, *Diamonds in the Rough: Corporate Paternalism and African Professionalism on the Mines of Colonial Angola, 1917–1975*

Carina E. Ray, *Crossing the Color Line: Race, Sex, and the Contested Politics of Colonialism in Ghana*

Sarah Van Beurden, *Authentically African: Arts and the Transnational Politics of Congolese Culture*

Giacomo Macola, *The Gun in Central Africa: A History of Technology and Politics*

Lynn Schler, *Nation on Board: Becoming Nigerian at Sea*

Julie MacArthur, *Cartography and the Political Imagination: Mapping Community in Colonial Kenya*

The Gun in Central Africa

A History of Technology and Politics

Giacomo Macola

OHIO UNIVERSITY PRESS ∽ ATHENS, OHIO

Ohio University Press, Athens, Ohio 45701
ohioswallow.com
© 2016 by Ohio University Press

To obtain permission to quote, reprint, or otherwise reproduce or distribute material
fromOhio University Press publications, please contact our rights and permissions
department at(740) 593–1154 or (740) 593–4536 (fax).
Printed in the United States of America

Ohio University Press books are printed on acid-free paper ⊗ ™

26 25 24 23 22 21 20 19 18 17 16   5 4 3 2 1

Library of Congress Cataloging-in-Publication Data
Names: Macola, Giacomo, author.
Title: The gun in central Africa : a history of technology and politics /
  Giacomo Macola.
Other titles: New African histories series.
Description: Athens, Ohio : Ohio University Press, 2016. | Series: New
  African histories
Identifiers: LCCN 2015046077| ISBN 9780821422113 (hc : alk. paper) | ISBN
  9780821422120 (pb : alk. paper) | ISBN 9780821445556 (pdf)
Subjects: LCSH: Firearms—Africa, Central—History. | Firearms—Social
  aspects—Africa, Central—History.
Classification: LCC U897.A352 M33 2016 | DDC 683.400967—dc23
LC record available at http://lccn.loc.gov/2015046077

FOR DAVINA

"We don't follow no crowd / They follow us"

Contents

Illustrations

Acknowledgments

Considering the increasingly marginal status of precolonial African historiography in the United Kingdom and elsewhere, the curiosity sparked by this book came as a pleasant surprise. I am indebted to the following Africanists for taking the trouble to comment on sections of the manuscript: David Birmingham, Paul la Hausse de Lalouvière, John Iliffe, Dirk Jaeger, Bill Nasson, Andrew Roberts, Ken Vickery, and, especially, Jean-Luc Vellut. Not only did I learn a great deal from Jean-Luc's own work on the subject of firearms in central Africa, but he was also generous enough to put me in touch with Paul Dubrunfaut, the supremely knowledgeable keeper of firearms at the Musée Royal de l'Armée, Brussels. Needless to say, none of these scholars ought to be held responsible for any errors and/or misconceptions that remain in the book despite their much appreciated cooperation.

Other colleagues contributed in less direct but still invaluable ways. Bill Storey kindly participated in a conference I co-organized in Canterbury in May 2011. Although our approaches to the history of firearms in Africa are far from identical, I readily admit to having been initially much influenced by his *Guns, Race and Power in Colonial South Africa*. Achim von Oppen, the author of an important and original study of the upper Zambezi and Kasai region, *Terms of Trade and Terms of Trust*, graciously allowed me to use one of his splendid photographs for the cover of this book. Ray Abrahams, Jeff Hoover, Martin Walsh, Judith Weik, and Samba Yonga helped me with their linguistic expertise. Hugh Macmillan, John McCracken, and Kings Phiri pointed me in the direction of fundamental sources in Scotland, where I also benefited from the hospitality of, and discussions with, Tom Molony. In Lubumbashi, Donatien Dibwe dia Mwembu and Léon Verbeek showed a keen interest in my project and, alongside Pierre Kalenga and Liévain Mwangal, went out of their way to facilitate it. Doing research in my adoptive country, Zambia, is a rather easier proposition than in the Congo. Not its least attractive feature is the extensive support network on which I am able to rely. Marja Hinfelaar, Bizeck J. Phiri, and Mauro Sanna have always been the most dependable of friends. Institutionally, both the National Archives of Zambia and the Livingstone Museum have invariably done their best to accommodate all of my research requirements. In Livingstone, special thanks must go to the then keeper of history, Friday Mufuzi, who, alongside Flexon Mizinga, the secretary of the Zambian National Museums Board, granted me permission to view and photograph some of the firearms held at the Livingstone Museum.

Few of my close personal friends are to be found within the (occasionally suffocating) walls of the academy. But, precisely because they are few, my academic comrades are all the more precious to me. My greatest debt goes to Harri Englund, upon whose friendship and unselfish readiness to offer advice I have always been able to count. Despite his busy schedule, Harri has always found time to comment on the various chunks of the manuscript that I mercilessly inflicted on him. I have, moreover, very fond memories of our short stint of joint fieldwork in Zambia's Eastern Province (my rabid envy of his proficiency in Chichewa notwithstanding). In Canterbury, Pratik Chakrabarti, Nandini Bhattacharya, Leonie James, Ambrogio Caiani, and Jackie Waller have been rocks of support, spoiling me with their hospitality and generally keeping me on the straight and narrow. Walima Kalusa and Joanna Lewis are both excellent historians and great mates; my knowledge of the central African past has been much enriched by our frequent, rambunctious conversations. Despite having to deal with personal tragedy, Jan-Bart Gewald has remained an exceptionally big-hearted friend over the last ten years or so. Jan-Bart and his late wife, Gertie, also played a critical role in drafting the research proposal that secured the funds without which this book could never have been written. Long may you run, settler boy! Exception made for his interest in ornithology (which, I am convinced, rather cramped our style in southern Congo), Robert Ross was a great traveling companion, whose curiosity and imaginativeness always kept me on my intellectual toes. Robert, too, gave the manuscript a careful reading and made a number of vital suggestions about how best to structure it. Ian Phimister and I have at least one thing in common: the feeling that forbearance might well be an overrated virtue. If this is not the basis of a solid friendship, then I don't know what is. Ian was also kind enough repeatedly to host me in his Bloemfontein lair, where parts of this book were first presented to wonderfully attentive audiences, and, later, prepared for publication. Other research seminars and conferences where I discussed my initial findings took place at SOAS (University of London), the University of Kent, and Leiden University. I am obliged to the attendees for their stimulating comments.

Having until recently been the only Africanist in the School of History of the University of Kent, the feeling of intellectual isolation is not unknown to me. I am therefore sincerely grateful to my PhD students, Jack Hogan, John Kegel, and Peter Nicholls, for having made my Kentish "Bantustan" a less lonely place. Besides learning to put up with my tactlessness and impatience, and producing an excellent set of maps, Jack also generously shared much useful primary material from his outstanding work on the abolition of slavery in western Zambia. Since 2012, my third-year special subject—"Kingdoms of the Savannah: The Political History of Central Africa, c. 1700 to c. 1900"—has attracted a number of terrific students. I am both gratified and touched by their readiness to be challenged by a comparatively recondite—and uniquely complicated—subject. For those among them who are contemplating Africanist careers, the message is simple: there's still plenty of room in the savanna.

This book was made possible by a three-year-long grant from the Nederlandse Organisatie voor Wetenschappelijk Onderzoek and the related concession of an extended study leave by the University of Kent in 2009–2011. Naturally, I am much indebted to both organizations. A very early version of chapter 2 has been published as "Reassessing the Significance of Firearms in Central Africa: The Case of North-Western Zambia to the 1920s" in the *Journal of African History*. Sections of chapters 5 and 6 are reprinted by permission of the publishers from "'They Disdain Firearms': The Relationship between Guns and the Ngoni of Eastern Zambia to the Early Twentieth Century," in *A Cultural History of Firearms in the Age of Empire*, which volume I had the pleasure of editing alongside Karen Jones and David Welch. Also, I am grateful to Gillian Berchowitz, the director of Ohio University Press, and the editors of its splendid New African Histories series. Jean Allman, Allen Isaacman, and Derek Peterson all took a keen interest in the project, waited patiently for the manuscript, and then offered perceptive remarks about how best to go about improving it.

My daughter Davina, around whom my world revolves, has been clamoring for a dedication for quite some time. Finally, here it is, bambina: *questo libro é tutto per te.*

A Note on Hereditary Titles

Hereditary titles are italicized throughout to distinguish them from personal names. I use standard roman type only when the title in question is accompanied by the name of its holder, or when the context makes it plain that I am alluding to one particular, if unnamed, individual incumbent. Thus, for example, I write the "Ruund of the *Mwant Yav*" to refer to the people who acknowledged the sway of an undetermined number of successive holders of the royal title (*Mwant Yav*), but the "Ruund of Mwant Yav Mukaz" to describe the followers of one specific king—in this case, Mukaz, who briefly held the reins of power in the Ruund heartland in the 1880s.

Firearms and the History of Technology in Africa

THE POLICE post in Bunkeya, Mwami Msiri's old imperial capital in the present Territoire de Lubudi of southern Katanga, Democratic Republic of the Congo ("DR Congo"), is a shabby—if colorful—place. On a pleasant day early in August 2011, Robert Ross, Pierre Kalenga, and I entered it to announce our presence in town for a brief stint of research. The plan was to bring to an end our dealings with the local representatives of the Agence Nationale des Renseignements (or "ANR," somewhat optimistically described to me as "the Congolese FBI") as soon as diplomacy and politeness made it possible. In the event, something caught our attention and made us stay longer than we envisaged: a heap of rusty firearms occupying a sizeable portion of the floor surface of the tiny room into which we had been ushered. The guns in question comprised a dizzying variety of models, though percussion-lock muzzle-loaders were the most numerous. What all of these firearms had in common, however, was that they had been manufactured locally, using gun scraps, homemade pieces, and industrial parts. Upon inquiry, we discovered that the guns—commonly going under the obviously onomatopoeic name of "*poupous*"—had been subject to precautionary confiscation from local residents in the spring of 1997, when Laurent D. Kabila's forces had entered the town during the campaign that would shortly thereafter result in the overthrow of long-serving dictator Mobutu Sese Seko.

"Surely they don't work now?" I asked.

"No, but many of them were already useless back in 1997," was the reply of one of the two officers.

FIGURE 0.1. Homemade poupous, Bunkeya, DR Congo. Photograph by the author, August 2011.

"Why keep them, then?" I inquired, rather unimaginatively.

"Who knows? Most villagers have them. It's part of being a man . . . a *père de famille*."

At one level, of course, Bunkeya's poupous—accessible tools of self-protection in a country where threats of violence and predation have often been part and parcel of the daily lives of ordinary people—encapsulate

the troubled postcolonial history of the Congo. At another, they illus-
trate the historical relation between the Yeke of Katanga and hunting, an activity with which the poupous have been closely associated
from the early decades of the twentieth century. Once central to the
workings of Msiri's warlord state and the livelihoods of the people of
the district, hunting continues to play a marginal, seasonal role in the
domestic economy of some Yeke households.

But, as the final remark of the ANR officer suggests, there is more
to guns than meets the eye, and their social role in southern Katanga—
and, as I will argue, elsewhere—cannot be reduced to their military
and economic functions. Although regional specialists have been slow
to acknowledge the phenomenon, a "surplus of meaning" has clearly
been inscribed upon this technological artifact.[1] Besides working as defensive and hunting tools, Bunkeya's poupous have also been endowed
with a host of less predictable symbolic attributes. As adumbrated by
my informant, in some contexts homemade guns were and are probably less valued as operating weapons than as markers of masculinity
and signs of patriarchal status and self-reliance. The reasons why one
bundle of cultural meanings prevailed over several different possible
combinations are eminently historical. That is, they become accessible
to historians only when they are appraised in the light of the specific
experiences and worldviews of the people concerned and the changes
they underwent across different historical frameworks. In the case of
the Yeke, the story would have to begin with their emergence as a gun-
rich, conquering elite in the middle decades of the nineteenth century,
the time of Katanga's direct incorporation into the long-distance trades
in ivory and slaves.

Novelists have long sensed that the power of objects extends well
beyond their immediate service functions. Thus, Joseph Bridau, one of
the *Comédie humaine*'s characters, lamented the passing of the golden
age of French aristocracy in the following terms: "The fan of the grande
dame is broken. . . . The fan is now used only for fanning. Once a thing
is nothing more than what it is, it's too useful to serve the cause of
luxury."[2] More than a hundred and fifty years later, historians of technology and material culture have come round to Balzac's intuition,
and the view is now widely shared that artifacts are polysemous; that
is, they embody different meanings and fulfill several purposes, both
simultaneously and diachronically. In this respect, Katangese guns are
not at all unique. But their physical attributes are less easily reducible

to a mere manifestation of the human tendency to endow objects and technologies with symbolic significance. Notwithstanding the disparaging assessment of the automatic rifle–carrying ANR officer, when the poupous first made their appearance in Bunkeya in the early twentieth century, they represented a triumphant marriage of local inventiveness and high user demand. The craftsmanship and eclecticism that they exhibit demand our attention, for they speak of long-drawn-out, locally rooted processes of technological engagement and domestication. These processes lie at the heart of this book, which approaches the trajectory of firearms in central Africa from a culturally sensitive perspective that embraces both the practical applications of guns and the set of values and meanings that they have been taken to encompass.

Focusing as it does on the nineteenth and early twentieth centuries — the early history of central Africa's entanglements with gun technology — the exercise is mired in complexity. Given the nature and limitations of the available source material, the holistic treatment of firearms that I advocate will sometimes remain more of an ideal towards which to strive than a tangible realization. But the current "foreshortening of African history" recently decried by Richard Reid makes the effort worthwhile.[3] This book, then, is driven by a double ambition, seeking both to make a stand against the increasing marginalization of African precolonial history in the academy and to take up David Edgerton's call to shift the study of technology away from its "historically familiar surroundings."[4] My two overarching aims, in fact, are closely interlaced, for one key strategy to rekindle scholarly interest in precolonial history is to establish a dialogue with theories and concepts originating from other disciplines and historical fields. It is to a quick discussion of these literatures that the next two sections of this introduction are dedicated.

SOCIETY AND TECHNOLOGY

There used to be a time in which the relation between technology and society was understood in simple unidirectional terms: technological progress was the work of exceptional individuals, who deployed their genius and scientific prowess to invent the artifacts that mechanically transformed society and drove it forward, towards ever-increasing levels of well-being and/or mastery over previously unharnessed forces of nature. In this reading, technological evolution possessed a kind of inner, implacable logic. The great contribution of "social construction of technology" (SCOT) approaches has been to complicate this linear model of development and to hand back to users of technology their historical role. Beginning with the work of Trevor Pinch and Wiebe Bijker, whose "manifesto" first appeared in the mid-1980s,[5] SCOT theorists showed that technologies were invariably the outcomes of compromises—compromises that called into question the inventors' ostensible isolation from society and politics and that highlighted the inanity of any attempt to distinguish "between a world of engineering on the one hand and a world of the social on the other."[6] In so doing, they began to bring to the fore what they termed the "interpretative flexibility" of technology: the fact that a given technological artifact is open to more than one understanding and that its applications, far from always being the predetermined outcome of the intentions of inventors, are often also the result of the choices and predilections of users. What SCOT illuminated, then, was the agency of users in shaping technological innovation—and, therefore, producers' strategies— by attributing both predictable and unanticipated functions to specific artifacts. The histories of technologies, in sum, reveal that the latter have frequently been "employed in ways quite different from those for which they were originally intended."[7]

However, as pointed out by Ronald Kline and, again, Pinch in a famous intervention, the SCOT paradigm did suffer from some "important weaknesses" in its early formulation.[8] Constructivist students of technology reconceptualized the inventor/user nexus, but did not quite explode it. As agents of technological change, users were rightly conceived of as belonging to "social groups," but only rarely did SCOT theorists engage with these same groups' internal composition and the dynamics of power that underlay them. The focus of this scholarship— as Gabrielle Hecht remarked—remained squarely on the "construction of technology," rather than "on the construction of culture or politics."[9] This atrophied picture of social relations (what Pinch and Bijker themselves referred to in passing as the "wider sociopolitical milieu"[10]) was accompanied by a narrow focus on the functional—as opposed to the symbolic—properties of technologies.[11]

It is at this level that consumption studies, an important branch of "material culture studies" in the UK,[12] have proved especially useful in shifting the field forward. By locating consumers in much broader networks of relations than did early constructivist students of technology, sociologists and anthropologists, in particular, have articulated "the importance of the sign value rather than the utility value of things."[13]

Objects, in this perspective, are "socially and culturally salient entities," which "change in defiance of their material stability" and which are endowed with expressive and symbolic attributes.[14] To put it differently, they provide a means of communication, an idiom through which to convey a variety of aims relating to individual and collective identities. The meanings conferred to commodities by consumers "express cultural categories and principles, cultivate ideals, create and sustain lifestyles, construct notions of the self, and create (and survive) social change."[15] Material things, students of consumption have established, are embedded in human social relations, which they help forge, consolidate, and even subvert.

This concern for the "material constitution of sociality" has shaped the recent work of historians of material culture and their important debate about the origins and workings of modern consumer society.[16] The power of things to construct identities and signify status is central to much of this scholarship—as attested, for instance, by Deborah Cohen's influential study of the interiors of middle-class homes in nineteenth-century Britain.[17] In Cohen's expert hands, the story of Victorian domestic possessions is the story of their transformation from signs of sinful worldliness to means of individual self-expression in the face of the homogenizing pressures of mass society.

Within science and technology studies, a marriage of sorts between the findings of SCOT and anthropological approaches to consumption has been effected by analyses that adopt so-called "domestication" perspectives. As deployed by Anne Laegran (and, before her, by Merete Lie and Knut Sørensen), the category of "domestication" serves to capture the essence of the process through which

> individual users, as well as collectives, negotiate the values and symbols of the technology while integrating it into the cultural setting. . . . Through domestication, technology changes as well as the user and, in the next step, the culture. More than within other constructivist theories on technology and users . . . the domestication perspective enables a thorough analysis of the users without relating directly to the design and manufacturing of the technology. It allows for redefinitions of practice and meanings even after the construction of the technology is closed from the producers' and designers' points of view.[18]

Rather than stressing the "closure mechanisms" through which the meanings of technologies are "stabilized" once and for all,[19] domestication approaches foreground a continuing process of user reinterpretation and re-innovation, and the coexistence of alternative understandings of a given artifact—over and above the hegemonic codes that might originally have been loaded into any such artifact by producers, advertisers, or any others likely to overdetermine meaning.

As shown by Jeremy Prestholdt, domestication perspectives are especially useful in examining situations of cross-cultural consumption.[20] Decoupling users from inventors and designers, domestication perspectives make it possible to study appropriation as a creative act in itself. This is a powerful tool in exploring the life of any object, but especially so when looking at how ostensibly peripheral societies use externally introduced technologies—such as firearms—for their own purposes, and imbue them with functions and meanings that do not always replicate those for which the objects in question had first been devised in their original, usually Western settings.

David Howes, who reads cross-cultural consumption through the lens of "creolization," articulates an essential dimension of the phenomenon.

> When one takes a closer look at the meanings and uses given to specific imported goods within specific "local contexts" or "realities," one often finds that the goods have been transformed, at least in part, in accordance with the values of the receiving culture. . . . What the concept of creolization highlights . . . is that goods always have to be contextualized (given meaning, inserted into particular social relationships) to be utilized, and there is no guarantee that the intention of the producer will be recognized, much less respected, by the consumer from another culture.[21]

Owing something to Marshall Sahlins's seminal *Islands of History* and its stress on "existing understandings of the cultural order,"[22] Howes's key intuition is not only that processes of functional remaking and symbolic reinscription do take place, but also that such processes of recontextualization are shaped by local sociocultural conditions and political interests—conditions and interests that the dynamics of appropriation themselves might subtly transform. In this sense, "domestication" and "creolization" are coterminous categories, for each emphasizes the contingent dimension of technology transfer and consumption, and the extent to which the latter activities are interwoven with preexisting circumstances and resources.[23]

THE HISTORY OF TECHNOLOGY AND
CONSUMPTION IN AFRICA

The history of technology in Africa has scarcely received the attention it deserves. Writing in 1983, Ralph Austen and Daniel Headrick bemoaned "the neglect of Africa by general historians of technology."[24] The situation over the past thirty years has not changed a great deal, as even a cursory glance at such specialist journals as *Technology and Culture* and *History and Technology* reveals. As a result, the field has until recently been almost completely unaffected by the paradigm shifts summarized above.

On the African continent, despite vivid displays of grassroots inventiveness and eclecticism in the sphere of everyday technology, technological determinism—the notion, that is, that society is the passive recipient of innovation, by which it is "determined"—has enjoyed a much longer lifespan than elsewhere. It is not coincidental that the most influential book on technology in Africa is still Headrick's *Tools of Empire*,[25] which remains standard reading in most undergraduate courses on imperialism and the history of science and technology.[26] In *Tools of Empire*, European imperial expansion in Africa and Asia in the nineteenth century is presented as the simple, automatic result of innovations in the fields of transport, armament, and medicine affecting, with unprecedented impact, non-Western societies. The manner in which colonial (or soon-to-be-colonial) subjects received, engaged with, appropriated, and sometimes subverted these same technologies falls outside the author's argument. Headrick's more recent work has remained, by and large, faithful to this original interpretative scheme: his latest *tour de force*—a catalogue of inventions, from early modern shipbuilding to twentieth-century air control—is revealingly entitled *Power over Peoples.*[27]

Studies of the co-construction of technology and society in Africa are not completely absent. The impetuous spread of new communication technologies, especially mobile telephony, over the past decade or so has given rise to a significant literature.[28] Only rarely, however, has this scholarship adopted a more than tokenistic historical perspective. Although there are happy exceptions to the rule—Brian Larkin's historically informed account of media consumption in Kano, Northern Nigeria, for instance, or the emphasis placed on processes of African appropriation in a recent collection devoted to the history of the motor vehicle[29]—the points stand that students of past technological change have given African users short shrift and that the latter's deep history of engagement with externally introduced artifacts remains poorly researched and understood. Writing about a large swathe of the colonial world in the twentieth century, David Arnold has recently pointed to our ignorance as to "what indigenes, rather than colonizers, made of new technologies" and how these same technologies "were locally received and adapted."[30] Valid as they are for the colonial period, Arnold's remarks are even more cogent in respect to precolonial Africa, to which the bulk of this book is dedicated.

Altogether more impressive have been the achievements of anthropologically oriented Africanist historians who have studied processes of commoditization without presenting the spread of consumer goods as the reprehensible indication of "global homogenization" and the erosion of "cultural differences."[31] Rather, the agency of Africans in forging the practices of their daily lives has been central to a scholarship that—in the words of Timothy Burke—has sought to foreground the collective and individual "acts of will and imagination, engagement and disinterest" that underlie the consumption of commodities.[32] In challenging dominant understandings of modern globalization as a purely Western-driven initiative, for instance, Prestholdt has illuminated the nonutilitarian dimensions of Zanzibari consumerism in the second half of the nineteenth century and the extent to which "global symbols," such as Western manufactures, were deployed in accordance with local norms and "in the service of local image-making practices."[33]

A number of social histories of such widespread consumer goods as imported alcohol and clothing have reached compatible conclusions and provided important insights into the "orientational functions" of consumption in nineteenth- and twentieth-century Africa and the workings of the cross-cultural domestication that accompanied it.[34] Dmitri van den Bersselaar's *The King of Drinks* is a good case in point.[35] This excellent social history of Dutch gin in West Africa is especially commendable for casting the spotlight both on African initiative and the chronological dimension of processes of commodity appropriation by southern Ghanaian and Nigerian consumers from the nineteenth century onwards. It was West African consumers—much more than foreign producers and advertisers and colonial policy makers—who were responsible for the paradoxical post–World War I metamorphosis of Dutch gin from "a mass consumer commodity, an iconic consumption item of modernity," to "a good with restricted, ritual circulation, an aspect of African 'traditional' culture, its use bound up with ritual and the authority of those who claim[ed] the sanction of custom."[36] The new local meaning bestowed upon gin may have appeared "wrong" in the eyes of European producers (who nonetheless benefited from it). In reality, it made perfect sense in the context of the commodity's increasing rarity *and* such preexisting cultural parameters as Akan color symbolism and notions of purity.[37] In other words, gin—like every other imported commodity—was always "likely to be incorporated into African consumptive patterns in ways that [made] sense in the context of existing yet continually changing world views, rather than according to the intended uses of the foreign producers."[38]

The history of clothes has revealed similar findings. In an important essay, Jean Comaroff describes, *inter alia*, the "host of imaginative possibilities" that the missionary-promoted spread of European apparel opened up for Tswana chiefs and, to a lesser extent, commoners from the early decades of the nineteenth century. The consumption of imported clothing throughout much of the century was infused with "local signs and values" and was shaped by indigenous social hierarchies and political interests. Particularly indicative of this "promiscuous syncretism" was the case of the Bakwena ruler, Sechele, who in 1860 commissioned a Western-style suit to be made out of leopard skin. Here, contrary to what even some of his own subjects assumed, the chief was not merely giving in to mimicry and emulation of encroaching Europeans. On the contrary, by combining the autochthonous symbolism of the leopard skin with the prestige-enhancing attributes of European dress, he was making an "effort to mediate the two exclusive systems of authority at war in his world, striving perhaps to fashion a power greater than the sum of its parts."[39] More generally, the centrality of "local circumstances" and "local fields of power" in the remaking of Western-style dress and the values inscribed in them is one of the main threads of *Fashioning Africa*, a fascinating collection whose editor, Jean Allman, reminds us that "the meanings of one particular item of clothing can be, and often are, completely transformed when moved across time and space. . . . While Western-style dress may have been 'foreign' in origin, its gendered, social, and political meanings were constructed locally. . . . In short, fashion may be a language spoken everywhere, but it is never a universal language."[40]

While highlighting the multiple possible outcomes of domestication/creolization processes, all of these various histories of externally introduced consumer goods in Africa point to the centrality of preexisting social, political, and economic structures in orienting patterns of engagement with a given commodity or technology as it moves across cultural contexts. As will be seen, my reading of the history of firearms in precolonial and early colonial central Africa takes this insight to heart.

THE HISTORIOGRAPHY OF FIREARMS IN AFRICA

The history of technology in Africa—as the previous section has begun to argue—is both comparatively undeveloped and still largely steeped in obsolete paradigms. Firearms represent only a partial exception to the general rule, for while much *has* been written about this "proverbial old chestnut,"[41] historians have rarely gone beyond describing the "role" of guns in warfare and seeking to assess the extent to which their introduction "impacted" on African societies, primarily by bringing about changes in military tactics and organization.[42] Early scholarly attention to firearms (aptly illustrated by two special issues of the *Journal of African History* devoted to the subject in 1971[43]) must be placed in the context of a more general preoccupation with the modalities of the Euro-African encounter at the end of the nineteenth century, a key concern of the first generations of professional historians of Africa. Whatever the reasons behind this early flurry of interest in firearms, the literature it spawned showed, in the words of Bill Storey, "little awareness of the dynamic relationship between society and technology."[44] This literature's deterministic underpinnings are both indisputable and understandable. Their survival into the present, however, is hardly justifiable, given the intellectual advances summarized so far. As pointed out by a perceptive scholar, the long shadow of technological determinism accounts for a deep-seated inability to think of African firearms as anything other than military or hunting tools.[45] As some recent outstanding work demonstrates, the study of African warfare (and, indeed, hunting) remains critically important,[46] but one of this book's central contentions is that only when less predictable patterns of gun usage are taken into account does it become possible to do justice to the full panoply of African understandings of guns in the precolonial and early colonial period.

Meanwhile, disregard for the social construction of technology and the role of African users as agents of re-innovation also accounts for the traction still enjoyed by arguments that either downplay the overall significance of imported weapons on account of their technical shortcomings (see chapter 2 for a fuller discussion), or, at best, state the impossibility of generalization—based on the fact, for example, that "guns were important in particular places at particular times . . . but equally there are times when the scholarly pursuit of the gun is at best a red herring."[47] Richard Reid is certainly correct in implying that guns elicited varied reactions in eastern and central Africa in the second half of the nineteenth century; what remains to be fully explained are the reasons why the outcomes of processes of technological engagement could diverge so dramatically.

Some efforts towards the adoption of constructivist perspectives in the study of guns in Africa have recently been carried out, although these, too, have suffered from a number of limitations that the present work seeks to overcome. In *Guns, Race, and Power in Colonial South Africa*, Storey sets out to extend the analysis of firearms beyond the confines of military history and to "examine the ways in which technology, politics, and society are mutually constituted." Unlike all the studies that preceded it, *Guns, Race, and Power in Colonial South Africa* rightly refuses to attribute "agency to guns," even as it teases out their "importance . . . for social and political change."[48] Yet Storey is principally concerned with interracial relations in a colonial context, and the book's overarching theme is the analysis of the extent to which successive debates about gun ownership and trade contributed to define notions of citizenship and hierarchies of race and power on the imperial frontier. Thus, while both the utility and sign value of guns to settler communities are explored in great depth, readers learn rather less about the ways in which Africans—both within and outside the Cape Colony—domesticated the new technology. Scattered here and there are indications that Africans—no less than settlers—attributed complex cultural meanings to firearms and deployed them for a variety of internal purposes. Storey, for instance, mentions in passing that Africans in the Transkei regarded firearms as insignia of masculinity and that the Sotho resisted disarmament because, by the 1870s, "guns had become linked to the authority of the chiefs."[49] But these insights are not systematically developed. Drawing mainly on official sources and settler newspapers, *Guns, Race, and Power in Colonial South Africa* cannot convey a full appreciation of the complexities of African socio-cultural structures. Because of this, the history of African-owned guns presented by Storey is still primarily a history of their service functions. The same is true of Jeff Ramsay's article-length study, which merely hints at the "significance of firearms as symbolic markers as well as material instruments" in nineteenth-century Botswana.[50]

A more rounded treatment of the subject could have been expected of Clapperton Mavhunga, whose history of Gonarezhou National Park, in southeastern Zimbabwe, over the past hundred and fifty years is explicitly presented as an attempt to "work at the intersection of Science and Technology Studies . . . and African Studies" by charting the "interactions of people, technology and nature."[51] However, despite a more "Afro-centric" focus than Storey's book, Mavhunga's work remains, at heart, an environmental history only occasionally lifted by constructivist perspectives. Mavhunga is certainly not unaware of the dialectical relationships between technology and gender identities,[52] but, even here, cultural issues are only tangentially addressed. While making the critical point that, "in the face of local village mobile workshops," "the European's instruments" sometimes "acquired uses neither the European designers nor the hunters had bargained for,"[53] Mavhunga's discussion of firearms hardly moves beyond the material aspects of technology transfer and the use value of imported weapons. This limiting approach is also characteristic of Mavhunga's earlier work.[54]

Although imperfectly executed, Mavhunga's central argument—that the study of precolonial Africa has something to offer to science and technology studies—remains valid nonetheless. I maintain that a focus on one specific technological artifact—in this case, firearms—can go some way towards winning the same argument. In many respects, guns in precolonial central Africa work for me as the bicycle does for David Arnold and Erich DeWald in the context of colonial India and Vietnam: as a comparatively accessible and originally exogenous technology whose rapid—though not universal—spread enables one "to observe the wide variety of social uses and cultural understandings to which it gave rise."[55] By the nineteenth century, as will be argued below and in chapter 1, the interior of central Africa encompassed an array of political and cultural systems. This heterogeneity offers ample scope for comparison and makes it possible to illuminate the extent to which different societies responded differently to the same kind of technology, a point that an exclusive focus on relatively homogeneous—if, of course, highly stratified—Western societies tends to obfuscate. Following from this is the emphasis on technological *disengagement* in the third part of the book. The rejection of a given technology is one aspect of "the agency of potential users" that has "remained largely unexplored in domestication approaches."[56] One of this book's objectives is to show that acts of willful resistance were no less socioculturally determined than strategies of adoption.

As with any subject of historical inquiry, the trajectory of firearms could have been tackled from a variety of standpoints. It is therefore important to spell out at the outset what this book does *not* set out to do. The dynamics of the global arms trade—the subject of some well-researched recent works[57]—fall outside the scope of this book, which is more concerned with the African endpoints of such international small arms transfer systems as came to full fruition in the nineteenth century. Related to this is the fact that this book does not seek to present a comprehensive quantitative analysis (though, when available, quantitative data are interspersed in the narrative). This is, first, because patchy import records from the relevant coastal entry points in Portuguese and Zanzibari hands do not embrace the entire firearms trade, much of which took the form of smuggling.[58] It is thus unlikely that significantly more precise figures will ever be arrived at than the nineteenth-century estimates already in circulation (to which reference will be made in due course). Second, even if complete and reliable import statistics were available, they would not cast any light on the distribution of firearms in the interior, for which we must rather rely on the eyewitness accounts of literate observers (about which more will be said below). But the most important reason for not embarking on a quantitative study is that raw numbers are a poor indicator of patterns of domestication. What really matters to me are the uses to which central African actors put their guns, and such uses—be they practical or symbolic, conventional or innovative, consistent or inconsistent with the intentions of producers and traders—cannot be inferred from numbers alone. Once more, I find myself in agreement with Arnold and DeWald. Commenting on the comparatively small number of bicycles imported into colonial India and Vietnam, they explain that "the importance of the bicycle can best be measured less in terms of 'global diffusion' . . . than of the way in which it became implicated in the lifestyles and work regimes of a significant section of the population, and was caught up in issues of race, class, and gender, and of national identity and colonial state power."[59] *Mutatis mutandis*, I am making the same point with regards to precolonial central Africa.

More controversially, perhaps, this book is only tangentially concerned with the realm of the supernatural. Partly, this is in reaction to the once liberating but now increasingly formulaic tendency to portray Africans as "viscerally" religious beings, either "empowered or oppressed," but never left unaffected, by "invisible forces."[60] It is also a result of my contention that spiritual appraisals of guns, though not infrequent, were not the key factor influencing central Africa's terms of engagement with the new technology in the precolonial and early colonial eras. These factors, I argue, are instead to be located in a much broader understanding of social structures—one which, of course, encompasses religious manifestations but is by no means confined to them. In the second half of the nineteenth century, Lozi monarchs bolstered their newly regained position and asserted the modernity and worthiness of the social order they dominated by centralizing the gun trade of the upper Zambezi floodplain into their own hands and by inserting firearms into royal symbolism. At precisely the same time, the Yeke of warlord Msiri put them at the service of an unprecedented, market-driven system of economic spoliation in southern Katanga. And while Chokwe and Luvale hunters were incorporating them into their societies as irreplaceable markers of masculinity and individually owned tools for the production of human and animal capital, guns were willfully resisted in eastern Zambia and Malawi by Ngoni fighters bent on scaling their regimental organization through the display of heroic honor in hand-to-hand combat. In North-Western Zambia, meanwhile, kiKaonde-speakers had begun to employ them as a polyvalent currency as well. To be sure, all of these disparate "worldly" uses and understandings of firearms, and the attested proficiency of Central African gun-menders—a central theme of the chapters included in part two of this book and one which is also briefly touched upon in important essays by Joseph Miller, Jean-Luc Vellut, and Maria Emilia Madeira Santos[61]—do not rule out the possibility that some Africans at least appealed to supernatural forces to account for the ultimate functioning of firearms and to enhance their lethality (for one artifact showing evidence of such an appeal, see chapter 2, figure 2.2). Nor do these worldly uses and understandings mean that guns were not deployed, just like other weapons, in religious ceremonies intended to obtain the blessing of ancestors—not least in the context of ritually empowered activities, such as hunting and warfare.[62] But they certainly suggest that exoticizing readings of the relationship between Africans and firearms do not tell the full story, or even the most important part of it. A stress on invisible entities and forces, moreover, runs the unintended risk of driving a wedge between technology and human initiative—which is precisely what this book sets out to avoid.

My refusal to analyze technology as an independent variable—and, more generally, to attribute "agency" (whether "primary" or, as per Alfred Gell, "secondary") to material things—may be questioned by scholars such as Nicole Boivin, who has recently argued that the very physicality of objects or technologies, their "materiality," grants them the power to "act as agents independently of people."[63] My sense is that, no matter how sophisticated, attempts inspired by Actor Network Theory to overcome socially constructivist positions invariably end up reintroducing forms of technological determinism—or even evolutionism!—through the back door. While I make no apology for clinging to the essence of what Boivin belittles as "humanistic and idealistic thought," I am also readily prepared to concede the well-taken point, that the emphasis on the processes through which society conditions technology has sometimes led us to lose sight of the equally "urgent task of understanding how technology concurrently shapes society."[64] This book seeks to avoid this pitfall by examining both the ways in which firearms were incorporated into existing sociocultural relationships and the ways in which such acts of vernacularization rebounded on, and led to change within, the same sociocultural settings. Thus, to use Ann Stahl's terminology, this book is not a "history of a material," but a "material history"—a history, that is, built on the premise that "bodily engagement with material worlds" is both an effect and a cause of the "social and ideational realm."[65] The simplest possible way of summarizing my philosophical standpoint is that while I am loath to efface the ontological difference between objects and people—or between technologies and social relations—I am willing to accept that these domains transform one another. While being fitted into contexts, technological objects contribute to the formation of the same contexts.[66]

The history of European hunting and its interactions with African practices and ecological knowledge in the age of empire has already been expertly told with reference to a number of localities.[67] The same is true of the relationship between gun ownership and settler identities in specific colonial contexts.[68] Neither of these two subjects (for the study of which abundant sources could have been mobilized) is thus central to my purposes. My interest, once again, lies in guns in African hands, and in how guns changed—and were changed by—different African societies in the late precolonial period and beyond.

The final caveat to be introduced at this stage is that this book is not a technical compendium. Gun enthusiasts and encyclopedists should steer well clear of it. Granted, an understanding of the technical properties of successive models (and perhaps even a modicum of what Otto Sibum calls "gestural," or experiential, "knowledge"[69]) is necessary meaningfully to write about them, but firearms as collectable objects, sporting tools, or aesthetic products are of no intrinsic interest to me. Guns, in my reading, are no more (and no less!) than a useful prism through which to examine some of the most significant and abiding aspects of the history of central Africa in the nineteenth and the twentieth centuries. One of my most sincere hopes is that my efforts, circumscribed and provisional as they are, might still go some way towards revitalizing engagement with a region of the continent and a period of its history that, despite having lain very close to the heart of the Africanist canon only a few decades ago, have lately suffered from serious scholarly neglect.

DEFINITIONS AND OVERVIEW

Drawing on a range of theoretical concepts originating from outside the field of African studies, this book offers the first detailed history of firearms in central Africa between the early nineteenth and the early twentieth century. Intended as an exploration of the intersections between technology, society, politics, and culture, it adopts a comparative perspective to chart, and account for, different user and potential user reactions to the same externally introduced technology.

All of the case studies presented in this volume belong to what might be loosely called the interior of central Africa—or, more precisely, the central portion of the southern savanna, the vast stretch of open grasslands and woodlands lying between the Congo basin rainforest and the Zambezi River, to the north and south, respectively, Lakes Tanganyika and Malawi, in the east, and the upper Zambezi and upper Kasai Rivers, in the west.[70] Over the past millennium, this macro-region of central Africa has been characterized by a blend of unity and diversity, and the interplay of continuity and transformation. Prevailing ecological conditions dictated the basic parameters of historical development for the Bantu-speaking colonists who made the area their home. Defining structural forces included the overall sparseness of population and the scarcity of the natural resources at its disposal. From the standpoint of Fernand Braudel's "geo-history," then, the central savanna should be viewed as the site of a centuries-long, unspectacular struggle on the part of farmers, fisherfolk, and, less commonly, cattle keepers to make the most of their harsh environment.

Social and economic "trends" more relevant to the subject and chronological framework of this book also reveal significant underlying commonalities in the historical experience of the peoples of the central savanna. By c. 1700, large-scale, centralizing kingdoms were being formed in comparatively favorable ecological locales, distinguishable from surrounding districts by the availability of either fertile alluvial soils or locally and regionally tradable resources, or both. By this stage, however, centers of dynastic power still resembled relatively isolated islands in a sea of micro-polities shaped by the "equalising pressure" of predominantly matrilineal descent rules and the fissiparous tendencies of village life.[71] At first, the significance of external commercial influences—of which firearms would eventually become a most fundamental by-product—was limited; when and where it did take place, historical change was still primarily the result of the playing out of endogenous forces. In the nineteenth century, however, the trade in such tropical commodities as ivory and slaves became more and more important to the political economy of the region. By the middle decades of the same century the central savanna had turned into a veritable commercial crossroads: the meeting point of two converging frontiers of long-distance trade anchored in the seaports of present-day Angola, on the one hand, and Tanzania and Mozambique, on the other. The compressed time frame within which the bulk of the central savanna came to be incorporated into global exchange networks is an additional reason for treating it as a discrete historical unit in the late precolonial period.

Within the broad framework of this shared historical experience, however, internal diversity remained salient. Indeed, it became sharper, because the peoples of the savanna responded differently to the challenges and opportunities ushered in by the advance of merchant capital. The nineteenth century in east-central Africa was no doubt traumatic, and the notion of "military revolution" has recently been deployed to describe the increasingly violent and militarized nature of politics in this era of long-distance trade.[72] Still, preexisting hierarchies and patterns of governance were not uniformly obliterated by the rise of "new men" and their openly predatory and entrepreneurial political formations.[73] Meanwhile, not all militarized new states owed their raison d'être to involvement in global commerce, and there remained numerous clusters of decentralized authority that avoided incorporation into expansive states—regardless of whether the latter were the heirs of time-honored political traditions or the products of new economic circumstances. Even at the height of international trade and political turmoil, the lives of a large number of central African peoples continued to be organized around small-scale sociopolitical structures.

My reliance on the category of "gun society" also calls for a brief introductory commentary. In this volume, the expression is used in the most general and loose possible sense: a gun society is one in which firearms are put to momentous productive, military, and/or other symbolic uses, over a sustained period of time and by a politically or numerically significant portion of the population. To be sure, a more analytically precise, Marxist-influenced definition could have been adopted, with gun societies being described as societies in which the majority of the available guns are utilized as tools of production—that is to say, as hunting implements or military weapons destined to secure both human and material booty. In the event, however, since one of the book's key objectives is precisely to foreground the variety of sociocultural—as opposed to narrowly military or economic—uses attributed to guns in the central African interior, a less restrictive definition was deemed more appropriate.[74]

The central savanna's diversity-in-unity opens up an exciting range of comparative possibilities for the historian interested in investigating conflicting local responses to the same kind of imported technology. This book thus contrasts such gun societies as existed on the upper Zambezi—the border area between present-day Zambia and Angola—and in Katanga, southern DR Congo, in the nineteenth century with communities—primarily the Ngoni of eastern Zambia and Malawi—characterized instead by processes of technological disengagement. Critical as it is, however, the dichotomy between adoption and rejection does not exhaust the history of firearms in the central savanna, for gun societies differed from one another in numerous important respects. The case studies presented in the second part of the book serve to underscore this point. Besides boasting sufficiently detailed sources, the upper Zambezi and Katanga regions comprised a range of political and cultural systems: from ancient monarchical societies to "stateless" ones, passing through new market-oriented warlord polities. These disparities were reflected in different patterns of gun domestication, for different were the configurations of preexisting sociopolitical interests with which the new technology interacted.

Chapter 1, a broad survey of the political and economic history of the central savanna in the eighteenth and nineteenth centuries, works as an overture. Aimed at the nonspecialist reader, it is intended to enable him/her to negotiate his/her way around the more specific stories of technological engagement—and disengagement—that follow it. It provides a sense of the workings of power and international trade in the macro-region with which the book is concerned, foregrounds the diversity-in-unity that characterized it, and introduces the theme of firearms and the various functions that they could be made to perform.

Chapters 2 and 3 chart the emergence of gun societies on the upper Zambezi and in parts of Katanga in the nineteenth century. Their principal contention is that firearms mattered more to the late precolonial history of these areas than existing studies are prepared to concede. The argument, however, is not couched in simple quantitative terms, not least because such an approach sidesteps the difficult question of the technical weaknesses of the hardware of violence that global trade was then making available to inland societies. Rather, in keeping with the book's theoretical framework, the two chapters contend that the diffusion and popularity of muskets in the two areas can best be understood by examining, first, the ways in which central African peoples learned to minimize the deficiencies of imported weapons, using them profitably for both economic and military purposes, and, second, the acts of domestication through which they infused the new technology with local meanings that were sometimes at variance with those that it had originally been assigned in the contexts of its production. This heterogeneous process of technological consumption, it will be shown, was in every instance informed by the social and political circumstances in which the imported technology was received.

Looking ahead, chapter 5 serves as a counterpoint to the book's second and third chapters. It discusses precolonial military conservatism among the Ngoni of Zambia and Malawi, who resisted the adoption of firearms for war purposes, as they regarded the new technology as corrupting and emasculating. Sociocultural opposition, here, had to do with the fact that firearms threatened hegemonic notions of masculinity and honor constructed around *combat à l'arme blanche.* In so doing, they also threatened to foreclose the opportunities for individual advancement inbuilt in Ngoni polities and their age-grade regimental systems.

The paradoxical outcomes of the imposition of colonial rule from the end of the nineteenth century are described in chapters 4 and 6. Gun laws in British Northern Rhodesia came eventually to be regarded as essential "pacification" tools, serving to symbolize the curtailment of African citizenship rights on which the edifice of European domination was predicated. They thus spelled the end of the gun-centered systems of social relationships that had dominated the upper Zambezi region throughout the second half of the nineteenth century. The case of the Yeke of southern Katanga was different, mainly on account of their close alliance with the Congo Free State and its armed forces during the conquest and initial exploitation of the area in the 1890s. The irony is that, in southern Katanga, where early colonial rule was violent and pervasive, African-owned guns ended up retaining a more central—though by no means unaltered—role than they did in comparatively lightly administered North-Western Rhodesia. Conversely, in both Malawi and, especially, eastern Zambia, the arrival of the Europeans, the military defeat they inflicted on the lightly armed Ngoni, and the enforced end of the latter's raiding economy brought about a marked (and, once more, paradoxical) ideological realignment. Local honor discourses and the military technologies around which they revolved impressed British policy makers, who construed the Ngoni of Zambia as a "martial race," partly on account of their enthusiasm for edged weapons and close combat. This led to large-scale recruitment of Ngoni into colonial paramilitary police forces. Under the new circumstances, the gun became everything it had not been in the precolonial context, gradually replacing the assegai as the central symbol of Ngoni masculinity and major vehicle for individual improvement.

The conclusion draws together these various themes and explores their contemporary relevance.

SOURCES AND METHODOLOGY

This book draws mainly on nineteenth-century travelogues, written records of oral tradition and literature, linguistic evidence, and early colonial material. As a commercial watershed and arena of sustained political and cultural exchanges, the central savanna attracted a considerable number of literate witnesses over the course of the nineteenth century. Their published and unpublished accounts vary in quality, often depending on the motives of the authors concerned. Thus, whereas full ethnographic descriptions are sometimes available (the works of such explorers as Antonio Gamitto, David Livingstone, and Henrique de Carvalho spring to mind in this context), other accounts—especially those of traders drawn to the region for primarily economic purposes—are more sketchy and less conducive to in-depth historical treatment.

These limitations, of course, are to be expected and form part of the daily staple of every historian. More intractable are the problems posed by the increasingly racialized context out of which the reports of nineteenth-century witnesses emerged. As documented by an extensive literature,[75] "Orientalist" biases permeate these sources, not least when they address the subject of non-Western warfare and its tools. Richard Reid has thus shown that on the eve of, and during, the "Scramble for Africa," African violence was often depicted as senseless, the result of savage passion rather than cold calculation.[76] These Victorian fantasies, of course, worked towards legitimizing the resort to extreme violence on the part of the Europeans themselves. In the context of the present discussion, however, the key point to be retained is that students of firearms must be fully alive to the extent to which many nineteenth-century Western observers felt inclined to belittle African life and ways of waging war, in general, and African marksmanship and knowledge of guns, in particular. The limited attention such observers devoted to matters of tactics is partly a consequence of this intellectual proclivity, and it explains this book's inability to offer more than perfunctory treatment of actual military maneuvers.[77]

To be sure, then, the historian of African firearms must come to grips with the sway of pervasive racially infused stereotypes in the available written sources. What needs to be avoided at all costs, however, is to throw away the baby with the bathwater. While accepting that Western travelogues are, to an extent, "discourses" that cannot be taken at face value, I, like Roy Bridges, maintain that text is not everything and that "the depiction of the 'Other' is, whatever the distortions, in some way related to what they were actually like."[78] Once more, the argument has been made most forcefully by Reid. European and other non-African observers of precolonial life and its conflicts "often reached the wrong overall conclusions . . . owing to the frameworks in which they were doing the business of observing and then writing . . . But they absorbed an enormous amount of what was going on around them, and understood a great deal more than they have frequently been given credit for."[79] Their accounts, moreover, never completely silenced African voices—just like their heroic descriptions of "lonely" itineraries though "uncharted wildernesses" never fully disguised their practical reliance on African intermediaries, skills, and manpower.[80] Much of the information that travelers recorded was derived from Africans. Though often rendered "virtually invisible," the African informant remains inscribed in the record, "his presence felt in much of the data and interpretation that frequently [was] posited as the author's own. Indeed, in the very texts that are held to be the clearest expressions of European prejudice, written by the harbingers of a new imperial order, we can also, if we listen carefully enough, hear a multitude of African voices."[81]

In sum, for all of the "racial, cultural and political shortcomings" of their authors,[82] the nineteenth-century accounts of traders, explorers, missionaries, and hunters continue to offer significant opportunities to the historian of late precolonial Africa. Their value to students of socio-economic change, in general, and technology, in particular, emerges with special clarity when these sources are weighed against oral ones. As pointed out by William Clarence-Smith several years ago, the key problem with the traditions of high political offices is their selectivity, which, in turn, is the direct effect of their "serving to reproduce the superstructures of a given society." Because of this, oral traditions—Clarence-Smith's neo-Annalist critique contended—hardly lent themselves to writing anything other than elite political history in the narrowest possible sense.[83] Although Jan Vansina was no doubt correct in replying that Clarence-Smith had overlooked the multiplicity of oral forms subsumed under the category of "oral tradition,"[84] the core of Clarence-Smith's argument holds more than a grain of truth: if political traditions were all there was, historians interested in precolonial social and economic dynamics would find themselves in a very tight corner indeed. Their task, moreover, is not made any easier by the realization that only limited trust can be placed in focused oral interviews centering on a period at several generations remove from the present. The few interviews that I carried out with renowned community historians in southern Katanga and eastern Zambia suggest that significant local historical knowledge continues to exist—especially, perhaps, about the modalities of the colonial encounter at the end of the nineteenth century. In only a handful of cases, however, was such knowledge independent of locally available published accounts and did it extend to the specific subject of this work.[85] Western travelogues allow the historian in some part to overcome such stringencies.

So, in theory at least, does historical linguistics (to which Vansina himself eventually turned his attention from the 1980s). Insofar as they embody evidence about the past, words are documents in their own right. The problem with our topic is that, not infrequently, guns were given onomatopoeic names (such as the *poupous* of Swahili-speaking southern Congo with which this introduction began) or drily descriptive ones (often semantically related to words for "fire," "noise," "smoke," or similar qualities). For obvious reasons, words of this type are scarcely conducive to historical treatment. Still, the vocabulary of gun societies does permit us to draw useful inferences. A particularly rich lexicon about firearms, for instance, is a sure indication of profound and intimate technological engagement, on the origin and nature of which some of the words in question might cast a specific light. Songs—as shown by numerous specialists[86]—are another important resource for the historian of precolonial central Africa, in general, and the student of firearms, in particular. Though published and unpublished collections of songs do not cover all the localities and societies I am interested in, the songs I do draw upon provide important windows into local cultural identities and economic practices, and the extent to which guns came to be entangled in both.[87]

Early missionary sources (mainly those produced by the Free Church of Scotland, active around Lake Malawi from the mid-1870s, the Paris Evangelical Missionary Society, whose representatives first visited the upper Zambezi in the late 1870s, and the Plymouth Brethren, in Katanga since 1886) and official and unofficial colonial records (from the terrifyingly explicit personal papers of Congo Free State official Clément "Nkulukulu" Brasseur to the more anodyne reports of the British administration in Northern Rhodesia) have been employed primarily to investigate aspects of the colonial encounter, the formulation and implementation of gun control laws, and early patterns of colonial police recruitment. When compiled by perceptive, ethnographically minded observers, however, these sources also illuminate at least some of the workings of the processes of technological domestication during the decades that preceded their authors' arrival on the central African scene.

Finally, of course, there is the information that can be extracted from surviving precolonial and early colonial guns themselves. Some of these weapons have remained in local hands. Others are preserved in both African and European museums. A few specimens appear in the photographs included in the book. As the relevant captions clarify, whenever possible, I availed myself of the opportunity to draw on expert technical knowledge and advice to "read" such material evidence.

In sum, the trajectory of firearms in central Africa in the nineteenth and early twentieth century was variegated and multistranded. So, too, must be the sources that permit us to study it.

PART I

Contexts

1 ⌐ Power and International Trade in the Savanna

THIS CHAPTER offers a preliminary overview of the main drivers of the history of the central African interior. Its principal aim is to contextualize the case studies presented in parts two and three of this book by exploring, first, the workings of power in the central savanna from c. 1700 and, second, the changes in governmentality precipitated by its growing involvement in global trading networks over the course of the nineteenth century. As Steven Feierman and other historians of eastern Africa have argued, such changes were often revolutionary, leading to the emergence of new social groups, new polities, and new ways of enforcing authority.[1] In the troubled nineteenth century, individual charisma and military success became undoubtedly more central to the wielding of political power than they had been in previous centuries. Still, the impact of violent innovation was not the same everywhere. The "hereditary" and "mystical" principles of political organization discussed in the first section of this chapter did not disappear overnight, and an exclusive stress on historical ruptures runs the risk of obfuscating patterns of continuity.[2] The experiences of dislocation and turmoil were pervasive, but so were attempts to neutralize or adapt to them. Commercially driven violence and the increasing availability of firearms could provide the bases for the growth of new warlord polities and related mercenary groups, and they could bring to a premature end preexisting state-building efforts. But they could also be harnessed by, and thus inject new life into, the latter. Broad generalizations, then, are not the best way to address the changing political culture of the central savanna in the nineteenth century. Neither should the one-sidedly gruesome descriptions that many coeval Western observers indulged in be swallowed hook, line, and sinker. What can be said with certainty is that the intrusion of merchant capital and its African spearheads left central Africa more politically and culturally heterogeneous than it had ever been at any time in its long past.

The chapter also introduces the theme of firearms, describing the timing and modalities of their arrival on the central savanna and offering some initial indications of the disparate reactions that they gave rise to. The interaction between the peoples of the central African interior and firearms must be regarded as an instance of cross-cultural technological consumption. African understandings of guns were as complex as they were contingent, and one of the overarching arguments of this book is that the meanings and functions that the peoples of the central savanna attributed to firearms were shaped by preexisting sociocultural relationships and political interests. Without an appreciation of the multiplicity and diversity of such relationships and interests, it is impossible to grasp the logic behind the heterogeneity in patterns of gun domestication that characterized the region. Guns, as later chapters will show, were appropriated differently by different groups, for different were the sociopolitical contexts into which the new technology came to be fitted.

The final objective of this chapter is to introduce nonspecialist readers to the intricacies of the precolonial history of a macro-region that is frequently overlooked in recent general syntheses. It is therefore unashamedly encyclopedic in tone and structure. Since it paints with a broad brush and covers a wide array of areas, peoples, and themes, the chapter might perhaps be regarded as a kind of *legenda*, to which readers might want occasionally to refer back as they proceed with the rest of the book.

THE CENTRAL SAVANNA TO THE EARLY NINETEENTH CENTURY

This section concentrates on the workings of power in the interior of central Africa before external trading influences began to make themselves uniformly felt in the nineteenth century. Any such discussion must begin by stressing that the vast stretch of open grasslands and woodlands found between the Congo basin rainforest and the Zambezi River offered an altogether unpropitious environment for political

entrepreneurs. In this region of "pedestrians and paddlers,"[3] state building involved the consolidation of structures and institutions that
brought together for regulatory and extractive purposes several descent
groups, the main units in central African political relations over the
past thousand years or more.[4] Commonly, the process was predicated
on the recognition of an overarching center of power providing a unify-
ing principle of hierarchy: a chief or a king holding a dynastic name or
title. The title was vested in a specific kin group, but its sway was also
acknowledged by other lineages, who were themselves the keepers of
subordinate titled positions and who might sometimes compete among
themselves for the topmost dignity. But this was easier said than done,
for the region's scattered population, vast and easily traversable spaces,
and relative scarcity of natural resources magnified the challenges of
state building. As John Darwin aptly put it, where "rebelling meant
no more than walking away to found a splinter community," the job of
leaders was very tough indeed.[5] In the central savanna, even more than
elsewhere in sub-Saharan Africa, the key objective of aspiring big men
and state builders was always to establish durable claims over the labor
and loyalty of unrelated people. Given the frequent absence of standing
armies until the latter part of the nineteenth century, violent conquest
was only effective in the short term and when it proceeded alongside
less disruptive, "softer" forms of rule. Different societies came up with
different solutions to overcome the parochialism of localized descent
groups. Invariably, such solutions were related to the ecological speci-
ficities of their respective areas.

An early center of political experimentation in the interior of central Africa was certainly located in the Upemba Depression. Archaeo-
logical evidence in the form of copper ornaments and small iron bells
suggests that processes of social and political differentiation were at
work in this comparatively densely populated floodplain on the upper
Lualaba River, in present-day southern Congo, since at least the first
centuries of the second millennium.[6] The rise of wealthy ruling groups
in the area may have had something to do with the need to contain inter-
lineage competition focusing on access to the floodplain's rich, but
finite, alluvial soils and to its game and fish resources. The same authorities might have also been responsible for coordinating such hydraulic
works as were required to keep local economic life viable.[7] Experiments in conflict management and political integration on the upper
Lualaba are likely to have influenced developments in its immediate

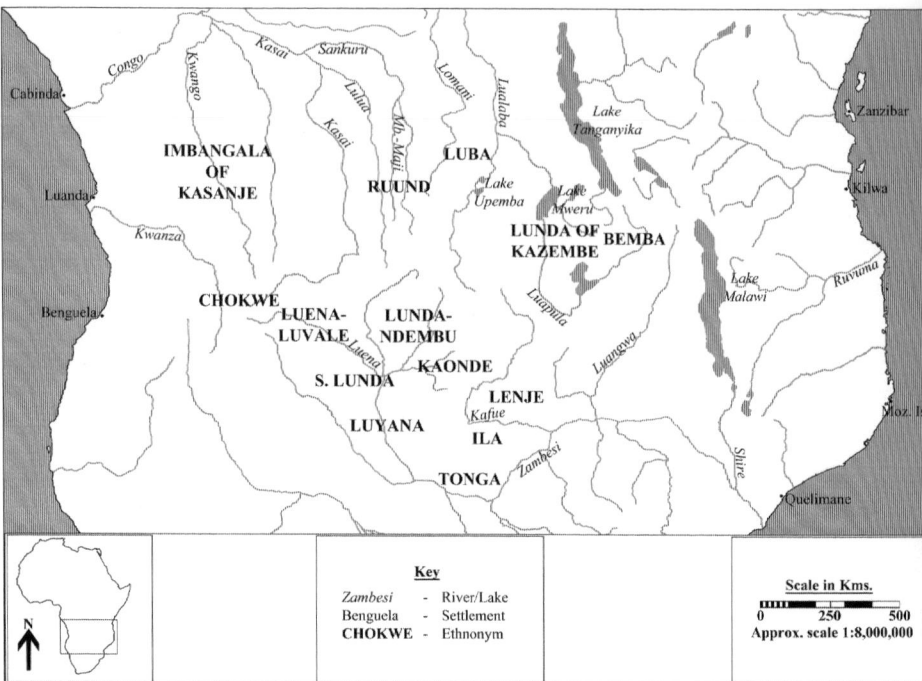

MAP 1.1 The central savanna, c. 1800.

surroundings, beginning with what would become the heartland of the Luba "Empire," the district located between the Lualaba and Lomani Rivers. Alternatively (or additionally), it is also possible to speculate that control over access to scarce trading resources—the salt and iron with which the future Luba core area was endowed—enabled one specific descent group to emerge as locally dominant and to regulate and tax the visits of outsiders seeking the same resources.[8]

By c. 1700, the time that a Luba dynastic kingdom becomes recognizable in the oral historical record,[9] elaborate political hierarchies, revolving around the *Mwant Yav* (*Mwata Yamvo*) royal title, had also come into being among the Ruund, to the southwest of the Luba.[10] The first substantial written mention of the Ruund state, by the Angolan slave trader Manoel Correia Leitão, dates to 1756. By then, the "Matayamvoa" was being described as a "powerful" conqueror and his followers as "terrestrial Eagles," raiding "countries so remote from their Fatherland only to lord it over other peoples."[11] Well-known traditions expounding on the marriage between the Ruund princess Ruwej and the wandering Luba hunter Chibind Yirung (Chibinda Ilunga) have frequently been interpreted as implying some form of Luba military conquest or, at a minimum, strong Luba influences on the genesis of the Ruund kingdom. In fact, the linguistic data examined by Jeff Hoover in the 1970s and the objective differences between the Luba and Ruund political systems suggest the playing out of more complex and longer processes of mutual borrowing than the conquest state model allows for.[12]

The twin institutions of positional succession and perpetual kinship—the Ruund trademark contribution to the precolonial political history of the central savanna—were certainly endogenous innovations.[13] Jan Vansina has recently called them "a stroke of genius."[14] Positional succession and perpetual kinship established permanent links between offices rather than individuals, and this meant that the Ruund kingdom "ideally consisted of a web of titled positions, linked in a hierarchy of perpetual kinship" and occupied by people of different background.[15] Because these institutions could be adopted without disrupting preexisting social structures, they became wonderfully effective means of imperial expansion. Subordinate hereditary positions could be created for real or honorary sons of a given Mwant Yav; their descendants—no matter who they were, or how far they lived from the Ruund heartland on the upper Mbuji-Mayi River—would continue to acknowledge the original connection, quite independent of the actual biological relationship that would eventually obtain between them and the successors of the Ruund king by whom the appointment had first been made. The integrating effects of positional succession and perpetual kinship were reinforced by another Ruund technique of rule: the recognition of the role of the "owners of the land." The distinction between "owners of the land" and "owners of the people" was rooted in the ancient political culture of the savanna, but the Ruund systematized it and broadened its application in the context of an imperial strategy. Both in Ruund and Ruund-influenced Lunda states, the leaders of autochthonous groupings were not eliminated or marginalized. Rather, they were granted important ritual prerogatives. Ruund and Lunda political rulers were the "owners of the people," dealing with the nitty-gritty of daily governance. But, though they were largely excluded from the sphere of temporal government, the "owners of the land" were still accorded a glorified position in the new dispensation. This was partly because they were believed to be in contact with the spirits of their ancestors, who exercised forms of supernatural authority over the districts they had first colonized.[16]

Thus, between the eighteenth and the nineteenth century, while the Luba sacred kings (*Mulopwes*) expanded their sway by favoring the accession of select peripheral lineage heads and incorporating them into the *bambudye*, a cross-cutting secret society which they controlled, the workings of positional succession and perpetual kinship helped bring into being a Lunda "Commonwealth." The "commonwealth"— a definition which Vansina advocates in preference to the more traditional one of "empire"[17]—consisted of a network of independent, though interconnected, polities. While their leaders claimed real or putative origins among the Ruund, and though they recognized the *Mwant Yavs* as the fountains of their prestige, these Lunda kingdoms were not ruled from a single center and did not form a single cohesive territory. The "commonwealth," in fact, consisted of a series of dominions ruled less by Ruund proper than by elites who had adopted Ruund symbols of rule and principles of political organization.

This Lunda sphere of influence—the extent and workings of which were first reported upon by the famous Angolan *pombeiro*, Pedro João Baptista, at the beginning of the nineteenth century—was crisscrossed by tributary and exchange networks and covered a large swathe of the central savanna. Its easternmost marches were occupied by the kingdom of Kazembe, founded as a result of the collapse of a Ruund colony on the Mukulweji River towards the end of the seventeenth century and the subsequent eastward migration of a heterogeneous group of "Ruundized" title holders.[18] In the west, the holders of the *Kinguri*, the royal title of the Imbangala kingdom of Kasanje, dominating the middle Kwango River since the seventeenth century, also claimed Ruund origins and—as will be seen below—became the *Mwant Yavs'* principal trading partners in the eighteenth century. In the south, smaller Ruund-inspired Lunda polities took roots along the Congo-Zambezi watershed. Early written evidence shows that "travelers" from the Lunda-Ndembu polity of the *Kanongesha* ("Canoguesa"), near the present-day border between Zambia, Angola, and Congo, were wont to take "tribute" to the *Mwant Yavs* in the 1800s.[19] There is no reason to believe that the southern Lunda of the *Shinde* and related titles, further to the south, would have behaved any differently.

Political change was not necessarily the result of diffusion or borrowing. Processes of state formation could be more insular and self-contained than in the Luba and Ruund/Lunda cases. The Luyana (later Lozi) state is a good case in point. Its rise owed very little to external influences, but was instead shaped by the complex politico-economic requirements of the upper Zambezi floodplain. Unlike the Upemba Depression (and much of the central savanna), the upper Zambezi floodplain could support cattle keeping. Yet, in other respects, the two ecosystems were comparable, for the Luyana heartland, too, consisted of a special environment that needed to be closely managed if its economic potential was to be fully realized. Annual floods compelled the people of the plain to build their villages, grow their crops, and herd their cattle on both natural and artificial mounds. At the height of the floods, however, even such mounds had to be temporarily abandoned and temporary residence taken up on the plain's forested margins. These conditions favored the development of a particularly centralized form of administration. Not only did the Luyana monarchs (*Litungas*) become "director[s] of the public works of the kingdom," but control over the allocation of scarce natural and man-made mounds also gave them an important means with which to buttress their power and position.[20]

The floodplain's large labor requirements were met by means of internal slavery—historically much more pervasive in Barotseland than anywhere else in the central savanna[21]—and through the *makolo* system. The makolo were military and, especially, labor units to which all the inhabitants of the floodplain belonged from birth.[22] They, too, worked as a means of royal centralization, for in the eighteenth century members of the royal family were seemingly replaced as makolo leaders by appointive officials. In time, the creation of new makolo and related headships became an exclusive royal prerogative. Thus, makolo leaders ended up forming what Mutumba Mainga calls a "bureaucratic aristocracy," whose elevated social status depended solely on its alliance with, and loyalty to, the kingship.[23] A system of territorial governorships overlapped the makolo. Like the heads of the latter, the officials in charge of specific districts were also selected on the basis of merit rather than birth. Their functions were probably mainly judicial, since it was the makolo chiefs who controlled people "for purposes of raising an army, collecting tribute and recruiting labour."[24] The marginalization of the members of the royal family to the advantage of the *Litungas* and their officialdom had one important consequence. Deprived of accessible outlets for their ambition, Luyana princes began fiercely to compete for the biggest prize of all: the position of *Litunga*. This was especially the case during interregna or following successful conspiracies. In this, the kingdom of the Luyana resembled that of the Luba, among whom royal heirs were similarly barred from assuming positions of territorial responsibility. In the nineteenth century, protracted civil strife would become a feature common to both polities.

All of these centralized organizations—the Luba and Ruund nuclear kingdoms, the "members states" of the Lunda "Commonwealth," and the Luyana polity—partook of a political economy in which rights over people, including bonds of political loyalty, were predicated on the transfer of material goods.[25] Thus, a key function of rulers was to act as redistributors of resources—be they internal resources accumulated through tribute or, increasingly as we shall see, imported commodities resulting from participation in long-distance trading networks. On the upper Zambezi, for instance, the people of Bulozi, the floodplain heartland of the Luyana state, required such forest products as wood for canoes and bark for ropes. They, of course, also needed to secure access to the higher lands to which their villages were moved when the floods made the plain uninhabitable. The residents of the forest, on the other hand, turned to the floodplain for fish, cattle, and milk. This set of converging interests meant that the Luyana kings were strategically placed at the center of networks of accumulation and redistribution that ensured the circulation of the products of complementary ecological and economic zones. By controlling and manipulating such networks, the Luyana *Litungas* brought into being webs of dependencies and obligations that structured their power and broadcast it beyond the circle of their immediate followers.

Imposing, outward-looking capitals were essential ingredients of the mystique that surrounded most savanna kingdoms. Veritable instruments of rule, they functioned as tangible embodiments of royal power intended to impress their rulers' subjects.[26] But the growth of these urban agglomerates was also clearly related to the tribute-gathering and redistributive functions fulfilled by the royal courts that they housed.[27] The Portuguese explorer Antonio Gamitto, for one, was very taken by the administrative sophistication that regulated social life in the capital of Mwata Kazembe IV Keleka in 1831–32. Going by his rough estimates, the population of what he deemed to be the "greatest town in Central Africa" is unlikely to have numbered less than ten thousand.[28]

As already implied, political power in the central savanna also had important religious dimensions. Its holders were deemed responsible for ensuring the health and fertility of the land by propitiating their (or other relevant) ancestral spirits. A rule of thumb is that the more centralized the state, the more developed kingly cults revolving around royal ancestors were likely to be. Following their installation, peripheral allies of the Luba *Mulopwes* received a number of special insignia. Paramount among these was the white powder prepared by royal spirit mediums. Through potent symbols such as these, subordinate lineage heads "participated in the aura of Luba sacral kingship," implicitly acknowledging "the superiority of that kingship over local concepts of chiefship venerated in their home villages."[29] As for the Luyana state, the position of royal ancestors in its cosmology explains why, in 1886, Litunga Lewanika went to great trouble to persuade the missionary François Coillard to pay homage to the grave of his predecessor, Litunga Mulambwa Santulu. As convincingly argued by Gwyn Prins, this step was deemed necessary to establish the monarchy's ritual superiority over the new arrivals, who, being "perceived by ordinary people . . . as magicians," posed a threat to a kingly power that was itself infused with mystical attributes.[30]

Contrary to what a host of modern scholars tend to imply, processes of ethnic consolidation were not solely a colonial phenomenon.[31] While precolonial central Africa was most definitely not inhabited by discrete, impermeable, and mutually hostile collectivities (the "tribes" of colonial parlance), the fact is undeniable that the roots and some at least of the building blocks of present-day ethnicities are to be found in the regionally uneven processes of political integration with which this chapter has so far concerned itself. Thus, as Andrew Roberts pointed out several years ago, to be Bemba in precolonial times meant not only to speak chiBemba, but also to consider oneself a subject of the holders of the *Chitimukulu*, the probably Luba-derived title that, after becoming the preserve of one single lineage of the Bena Ngandu royal clan late in the eighteenth century, embarked on a process of sustained territorial expansion, gradually imposing its sway over much of present-day northeastern Zambia in the following century.[32]

Centralizing, hierarchical state systems had thus undoubtedly come into being in the central savanna by the eighteenth century. Still, away from hubs of dynastic power, the lives of most central African peoples revolved around more fragmented, smaller-scale sociopolitical structures. On the frontiers of emerging state formations, political life was confined within the boundaries of the village, or, as in the case of the "stateless" Tonga of present-day southern Zambia, the ritual territory of rain priests. Alternatively, it gravitated around competing chiefly titles, the relationships between which were fluid and subject to frequent renegotiation. In the closing decades of the eighteenth century, this latter kind of political landscape was characteristic of the Chokwe and Luvale (or Luena) of the upper Zambezi and Kasai Rivers, and of the kiKaonde-speaking sub-clans to their east.

Lacking overarching centers of power, some of these communities found it problematic—or unnecessary—to mobilize resources on a large scale. This accounts for the comparative slowness with which they reacted to long-distance trading opportunities, or the fact that, when coastal traders did appear on the scene in the nineteenth century, small-scale societies—such as, for instance, the Ila and Lenje along the Kafue River—could end up being regarded as reservoirs of slaves to be raided at will. In contrary cases, however, well-developed hunting and forest-harvesting economies could actually facilitate and expedite involvement in the market economy. Thus it was that, beginning in the late eighteenth century, the highly mobile Luvale and Chokwe spawned aggressive market-oriented dynasties (as would, some decades later, the Kaonde) that were ready to take advantage of the new circumstances ushered in by the expansion of the Angolan slave and ivory trading frontier. Their polities, however, never entirely lost their original traits, consisting of unstable coalitions rather than centralizing kingdoms. Near the sources of the Zambezi, for instance, competition between the *Chinyama*, the *Kakenge*, and other Luvale titles remained an important factor throughout the nineteenth century.[33] Among the Luvale and similarly mobile groups, power—as a perceptive visiting missionary observed in 1895—was more "diffused" than in settled monarchical communities, such as the neighboring Lozi (the Luyana's successor state).[34] Partly because of this, on the frontiers of large-scale state systems, ethnic cultures did not coincide with coherent political unities, and the influence of the colonial "creation of tribalism" *would* be profound and transformative.[35]

THE IMPACT OF MERCHANT CAPITAL

The incorporation of the central savanna into globalizing trading networks in the nineteenth century was bound to have serious repercussions on the political landscape and patterns of governance described above. The increasingly pervasive impact of the outside world and the complex changes it gave rise to form the subject of this section.

Merchant capital and its African representatives followed two broad axes of penetration into the central interior of the continent. Their convergence in the middle decades of the century turned the central savanna into an arena of accelerated economic and political exchanges. To understand how this came about, the development of both eastern and western itineraries and the activities of their chief protagonists need to be discussed in some detail.

The Indian Ocean Trading Frontier

Until the closing years of the eighteenth century, much of eastern and central Africa had been relatively isolated from the Indian Ocean coast, where Zanzibar was then emerging as a major commercial entrepôt under the leadership of the Bu Sa'idi dynasty of Oman. Indirect trading links, based on local and regional networks of exchanges, had certainly been in place, but these had not been of such magnitude as to transform the sociopolitical landscape of the interior. From the end of the eighteenth century, however, a growing international demand for east-central Africa's ivory and slaves resulted in the forces of commerce gathering unprecedented momentum. During the first few decades of the nineteenth century, the bulk of the long-distance trade through present-day central Tanzania remained in the hands of African caravaneers—most notably the Nyamwezi and related groups, such as the Sumbwa (see map 1.2). It is probable that by the 1800s some Nyamwezi caravans had already begun trading with the kingdom of Kazembe, to which they were attracted not only by ivory and slaves, but also by the copper that the *Mwata Kazembes* received as tribute from their subordinates in southern Katanga.[36] Direct Nyamwezi trade with Katanga itself dates to the 1830s or so.

As the century progressed, better capitalized and better armed coastal merchants—normally referred to as Arab-Swahili, that is, residents of Zanzibar or the Islamized Swahili towns under its loose sway—began to encroach on Nyamwezi routes.[37] One of the effects of this competition was to increase the destructiveness of the trade and to push its frontiers further and further inland—a development facilitated by the establishment of Zanzibari bases in Unyamwezi country and at Ujiji, on the eastern shore of Lake Tanganyika, in the 1840s. By then, as attested by Gamitto in 1831–32, coastal merchants had already begun to visit the Lunda of the *Mwata Kazembe*, meeting part of the royal demand for cloth and other "ostentatious" possessions, including

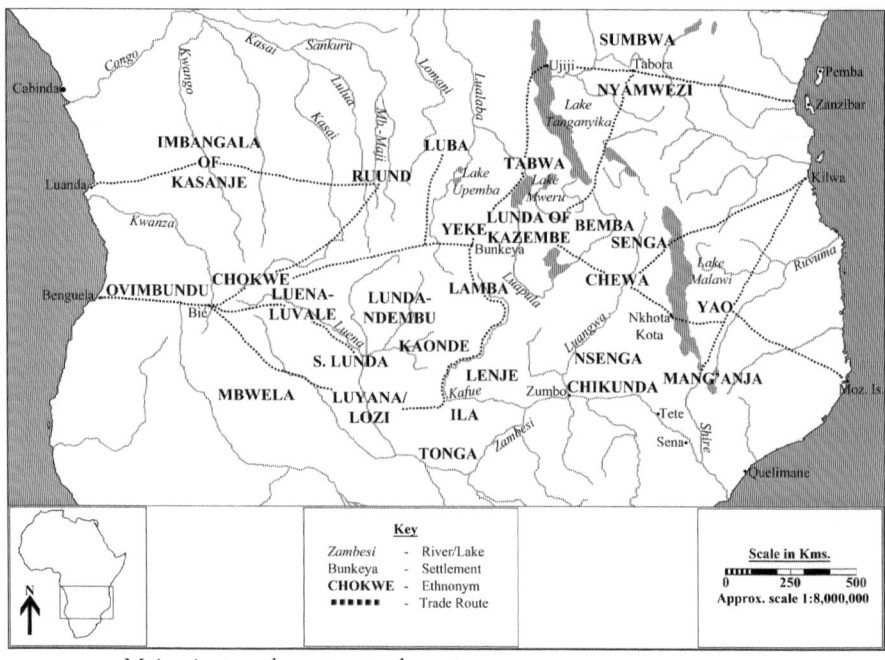

MAP 1.2 Main nineteenth-century trade routes.

"forty very clean shot guns, and six hunting carbines wrapped in lace-trimmed cloth."[38] Initially clearly deployed solely as charismatic symbols of royal wealth, the guns imported by Mwata Kazembe IV Keleka ("Chareka") eventually left his *Wunderkammern* (cabinet of wonders): c. 1850, Keleka's subjects—the Zanzibari trader Said ibn Habib reported—were "armed with muskets," some of which must have been used for military purposes.[39] These muskets were almost certainly flintlocks—either cast-off "Brown Besses" or trade guns modeled on the latter that Zanzibar was then beginning to ship to the mainland in sizeable quantities. It has been estimated that the number of guns passing through Zanzibar rose from about five thousand per year in the 1840s to nearly a hundred thousand in the 1880s, by which time the Bu Sa'idi capital had become one of the "main conduits for small arms transfers" in the entire western Indian Ocean.[40] After visiting the lower Luapula Valley, Said ibn Habib proceeded to the "great copper mines" of Katanga—where "a great many people [were] employed" and whose product was "taken for sale all over the country"—and the "petty states" of the Ila ("Boira") and Lenje ("Warengeh").[41] It was on the middle Kafue that Said ibn Habib first came across a party of long-distance traders from the far west: the Indian Ocean and Atlantic trading frontiers had intersected.[42] Taken together, they were set to overlay the bulk of the central savanna.

Further to the south, the ivory trade between Lake Malawi and the Portuguese colony on Mozambique Island, in northern Mozambique, had been in the hands of the Yao, then living predominantly to the south of the Ruvuma River and along the eastern shore of Lake Malawi, since c. 1700. In the latter part of the eighteenth century, as conditions of trade deteriorated on Mozambique Island, the Yao reoriented their activities towards the coast of southern Tanzania and, specifically, the ancient port of Kilwa, which had accepted Bu Sa'idi overrule in the 1740s. It was from about this time that the Yao, besides ivory, also began to deal extensively in slaves, a good number of whom were now acquired by French traders in Kilwa and imported into their Indian Ocean island colonies of Mauritius and Réunion. The Yao obtained some of their ivory from the Bisa, another group of semiprofessional traders who acted as intermediaries between the Lake Malawi region and political brokers further inland, such as, once again, the *Mwata Kazembes* of the lower Luapula Valley, with which the Bisa had been in contact since shortly after the foundation of the kingdom in about 1740. Eventually, in the middle part of the nineteenth century, the Bisa began to travel themselves all the way to Kilwa, where they offered their ivory at a cheaper price than the Yao did. Bisa competition in the ivory market—as Edward Alpers has suggested—further intensified Yao involvement in the slave business. The search for fresh supplies of slaves—many of whom were now deployed internally, on Arab- and Indian-owned plantations on the islands of Zanzibar and Pemba, and the Tanzanian coast opposite them—eventually led rival Yao gunmen to encroach *manu militari* upon Mang'anja territory along the southern shores of Lake Malawi and the heavily populated Shire Highlands of southern Malawi in the 1850s and 1860s.[43]

As in the Nyamwezi case, the Yao sector, too, witnessed increasing coastal activity as the century progressed. By the middle of the nineteenth century, an Arab-Swahili base had been established at Nkhota Kota, among the Chewa near the southwestern shore of Lake Malawi.[44] In the late 1870s, a smaller coastal settlement also sprang up among the Senga of the upper Luangwa River. Both sets of traders fed the traffic in slaves along the Malawi-Kilwa caravan routes, where violence—as Livingstone famously reported—had become a pervasive fact of life from at least the 1860s.[45]

Political relationships between Arab-Swahili traders and preexisting African authorities can be roughly subdivided into two phases. Since the vertical distribution of resources was an essential ingredient of their power, savanna kings and chiefs normally did their best to ensure that foreign trade took place under monopolistic conditions, or something that closely approximated them. Initially, given their precarious position, trailblazing coastal traders had little choice but to cooperate with domineering African authorities. It was only during a second phase, roughly beginning in the 1850s–1860s, that the Arab-Swahili resolutely moved against central African trading monopolies. This change of strategy was dictated both by their newfound military strength in the interior and by the increasingly competitive nature of the long-distance trading environment. The assumption of an explicitly political and military role on the part of coastal traders followed thereupon.

A watershed moment in the history of the region to the west of Lakes Tanganyika and Malawi was the defeat of the important Tabwa leader Chipioka Nsama III in 1867.[46] Chipioka, who some twenty years earlier had scattered an Arab-Swahili party, was attacked by Tippu Tip, the notorious Zanzibari trader and empire-builder, and his comparatively small but heavily armed following. In the encounter, Tabwa bowmen "died like birds! When the guns went off, two hundred were killed instantly and others were trampled to death! They fled. In one hour, more than a thousand died. Our casualties were only two slaves killed and two wounded."[47] Following Nsama III's defeat, his country entered a state of continuous civil war in which the Arab-Swahili—now stably settled between Lakes Tanganyika and Mweru—repeatedly played the role of kingmakers.[48] After the Tabwa, it was the eastern Lunda's turn to bear the brunt of what Marcia Wright and Peter Lary aptly termed the "Swahili version of British gun-boat diplomacy."[49] Controlled trade on the lower Luapula came to an end between the 1860s and 1870s, as coastal entrepreneurs worked systematically towards undermining royal monopolies and inaugurated an era characterized by successive foreign military interventions in the internal affairs of the much weakened kingdom.[50]

The Atlantic Ocean Trading Frontier

Having examined the chronology and prime movers of the Indian Ocean trading frontier, let us now turn our attention to the network of trade routes that developed in the opposite direction: from west to east.

Since the seventeenth century, the Portuguese coastal towns of Luanda and Benguela had been the main Angolan termini of a thriving slave trade. The itineraries flowing out of Benguela are especially important for the regions with which the next three chapters of this book are concerned.[51] Throughout the eighteenth century, most of the slaves exported via Benguela came from its immediate hinterland and the Umbundu-speaking, politically divided, central Angolan highlands. In the second half of the century, Luso-African traders settled on the plateau, especially in the kingdoms of Bihe (Viye) and Mbailundu. It was these merchants and independent local operators who extended the Benguela trading complex in an easterly direction, eventually reaching the upper Zambezi River in the closing decades of the eighteenth century.[52] At this time, the Luyana—as Livingstone was later to learn[53]— refused to export the slaves on whose labor the political economy of their core area depended. More willing commercial partners were found among Chokwe and Luvale hunters, who quickly learned to minimize the technical deficiencies of the mainly Belgian-made trade muskets imported by Bihean and other traders, to appreciate their versatility as tools for the production of animal capital and human booty, and to draw on their gendered symbolism. The direct incorporation of the upper Zambezi into the Atlantic economy went hand in hand with the spread of both social insecurity and new opportunities for marketing the region's primary products. These factors worked towards consolidating the sway of competing merchant chiefs. Internally divided Luvale trading principalities came into their own at a somewhat earlier date than the aforementioned Yao merchant dynasties in southern Malawi, but they obeyed similar market-driven logics and appropriated guns in similar fashion: as means of production, territorial expansion, and masculine affirmation. Slaves remained the upper Zambezi's dominant export until at least the 1850s, though the rise in ivory prices at the coast from the mid-1830s meant that ivory (and wax) exports gradually grew in importance. Later, during the 1870s and 1880s, they would be joined by natural rubber.[54]

In the mid-nineteenth century, Umbundu-speaking traders were probably the main carriers of the Angolan slave and ivory trade.[55] By then, their reach extended well beyond the upper Zambezi area. Having established solid relationships with the Kololo—Sotho migrants who temporarily overran the Luyana state in the middle decades of the century—the Ovimbundu trading (and, when circumstances permitted, raiding) sphere embraced the Ila and Lenje of the middle Kafue River and, eventually, the neighboring Kaonde of present-day Solwezi and Kasempa districts. Ovimbundu caravans—sometimes comprising several thousand free and enslaved porters (see figure 4.1)—also became very active in southern Congo. Since c. 1700, slave-dealing Ruund *Mwant Yavs* had been in contact with the Portuguese capital of Luanda through the mediation of the Imbangala of Kasanje.[56] For about a century, the rulers of Kasanje had been able to protect their middleman position. The bottleneck, however, had exploded early in the nineteenth century, as a result of the crisis of Kasanje.[57] From that point onwards, the Ruund capitals on the upper Mbuji-Mayi became a key destination for both Luso-Africans from the hinterland of Luanda and Ovimbundu from the Angolan plateau.[58] With the appearance of new actors on the scene, trade became less regimented, and the *Mwant Yavs* found it increasingly problematic to enforce such monopoly over foreign exchanges as they had enjoyed in earlier decades.[59]

Later, the predicament of Ruund royals was compounded by the arrival of Chokwe migrants. Beginning as elephant hunters and, later, rubber gatherers, the Chokwe moved down the Kasai Valley and settled near the Ruund heartland.[60] Historically, the Ruund had been poorly disposed towards firearms. In the eighteenth century, according to Leitão, they had regarded them as a "handicap to valor."[61] One century later, guns were still scarce among the nuclear Ruund.[62] This explains the ease with which musket-wielding Chokwe invaders carved out a dominant role for themselves during the internecine wars that marred the political life of the Ruund state after the death of Mwant Yav Muteb in 1873. By 1887, the year in which Carvalho visited the capital of a much enfeebled Mwant Yav Mukaz, the Ruund state was a shadow of its former self. Mudib, Mukaz's predecessor, had been killed by his erstwhile Chokwe backers and his capital destroyed. On that occasion, Chokwe warlords had captured "more than 6,000 people"—adults, children, and, especially, women, whom they incorporated into their expanding matrilineages.[63] By the late 1880s, Chokwe raiders were the real masters of the Ruund heartland and had initiated a phase of indiscriminate slaving, "leaving a virtual desert where the Lunda empire had once stood."[64] The southern members of the Lunda "Commonwealth," too, struggled to adjust to the conditions of the frontier market economy. As will be further seen in the next chapter, the southern Lunda of the *Shinde* and others were preyed upon by heavily armed Luvale slavers throughout the better part of the nineteenth century.

The effects of international trade on the central Luba state were, if possible, even more pernicious. This was largely a question of timing. Since they had remained more or less unaffected by the long-distance trade until the 1870s, the Luba only experienced it in its most disruptive, late-nineteenth-century guise. From about 1870, the Ovimbundu had become the main trading partners of Garenganze, Msiri's newly formed warlord state, the brutally extractive methods of which were then producing unprecedentedly large quantities of ivory and slaves for export. The Luba fish was hooked as a by-product of the commerce between Angola and Garenganze. Exploiting Luba internal divisions and lack of familiarity with foreign trade, Luso-Africans and Ovimbundu quickly established themselves as the dominant powerbrokers between the Lualaba and the Lomani Rivers. The civil wars that accompanied each royal succession became especially destructive, with different Luba factions drawing on the support of competing Angolan entrepreneurs. The early stages of this spiral of violence—one which would eventually result in the dissolution of the old Luba state as a cohesive territorial and political entity—were witnessed by Cameron in the mid-1870s. At the time of Cameron's passage through the Luba heartland, the followers of Alvez—one of the traders with whom Mulopwe Kasongo Kalombo had allied himself with a view to defeating his many internal opponents—were given license to plunder the most vulnerable of the king's subjects. "Any cultivated spot they at once fell on like a swarm of locusts, and, throwing down their loads, rooted up ground-nuts and sweet-potatoes, and laid waste fields of unripe corn, out of sheer wantonness." Cameron was certain that, "had they not been armed with guns," Alvez's followers "would never have dared to act thus, for on entering countries where the people carried firearms these truculent ruffians became mild as sucking doves."[65]

Between Political Innovation and Continuity

The interaction between the forces of global commerce and preexisting authorities forms one of the master themes of the history of the central savanna in the nineteenth century. The political effects of this increasingly close connection with the outside world, as I am presently going to argue, were profoundly ambivalent and contradictory. To be sure—as has already been noted—central African ruling elites often struggled to control the dynamics unleashed by the onset of the long-distance trade in ivory and slaves and the militarization of social relationships that it precipitated. Among the central Luba and Ruund and in most Lunda states, the erosion of royal monopolies over foreign commerce and the distribution of imported commodities led to fragmentation, increased violence, and enhanced social differentiation and levels of slave exploitation.[66] Under these circumstances, the growth of international trade resulted in the weakening of long-established elites, whose dominance was now challenged by the rise of "new men,"[67] often initially installed on the already contracting peripheries of the old political formations.

In a number of cases, these efforts at state-building in a period of widespread turmoil were entirely shorn of traditional legitimacy. The aforementioned Yeke state of Msiri, the latter a Sumbwa caravan leader turned empire-builder in the region lying between the collapsing Luba and Ruund states and the shrinking eastern Lunda sphere of control, is an especially clear example of the spread of warlordism in the central savanna. Better equipped to face the trials of the era of large-scale trading than were the old Luba and Ruund/Lunda aristocracies, Msiri and numerous other political opportunists—not least Lusinga, in northern Katanga, whose tragic story has recently been masterly told by Allen Roberts[68]—gave birth to violently entrepreneurial conquest states, sometimes ephemeral, sometimes more enduring. In these new political organizations, power rested less on religious sanction, heredity, and redistribution than on their leaders' personal achievements, successful involvement in commerce, and preparedness to resort to sheer violence to achieve their aims. The diffusion of firearms, and their recasting into primary means of military and economic domination, were often central to these processes of political realignment, which also drew part of their impetus from the related emergence of semiprofessional and cosmopolitan standing armies of brutalized young men.[69]

Although these developments were especially sudden in the interior of central Africa in this era of long-distance trade, they were by no means unique to it. The coastal societies of Angola, large-scale exporters of slaves from a very early period, had already witnessed the overthrow of established authorities and the rise of heavily militarized polities from the seventeenth century onwards.[70] Northeastern Tanzania offers a more proximate example. There, control of rain medicine and kinship relations, and the redistribution of internal tribute in livestock and labor, were the most important weapons in the political arsenal of the Kilindi rulers of Shambaa. In the closing decades of the nineteenth century, however, these ways of relating rulers to people were largely superseded by a new political culture. Revolving around Semboja, chief of Mazinde, closer to long-distance routes than Shambaai, the mountain heartland of the kingdom, such culture drew on trading connections and the slave-gun cycle to pose an ultimately unanswerable challenge to old Kilindi politics. A drawn-out civil war and a "complete victory for the forces of decentralization" were the end result. In Shambaa political thought, kingly power had always been "ambivalent. It could be used to bring life or to bring death." In the Pangani Valley in the third quarter of the nineteenth century, death became dominant, as the monarchical order ceased to be the guarantee of fertility, unity, and stability. Political leaders began to systematically prey upon their subjects, who were often left with no choice but to seek some degree of protection by joining the ranks of the slave raiders and warlords themselves.[71]

Still, apocalyptic descriptions of the central savanna on the eve of the European conquest should be rejected.[72] Revolutionary transformation was not on the cards everywhere, and the continuities in political tradition should not be underestimated. After all, as Ian Phimister has pointed out with reference to late nineteenth-century Zimbabwe, merchant capital sometimes "[modifies] existing social relations without decisively altering them."[73] Centrifugal forces and/or the onslaught of the new entrepreneurs of violence *were* sometimes effectively resisted. When this was the case, preexisting forms of authority and governance could actually be strengthened by participation in long-distance trading networks. The Bemba and, especially, the Lozi polities exemplify such processes. Unlike the neighboring Tabwa and eastern Lunda, the Bemba kept both the Arab-Swahili of Itabwa and the so-called "Senga Arabs" at bay. As a result, while the overall cohesiveness of the Bemba "federation" might not have increased greatly, each Bena Ngandu chief, dealing on his own terms with coastal traders, used imported commodities to augment his "ability to attract and reward supporters."[74] As will be seen in the next chapter, after overthrowing the Kololo in the 1864, the Lozi maintained their erstwhile conquerors' open-door policy vis-à-vis foreign traders, but reverted to being a slave-importing—as opposed to slave-exporting—society. This, and Litungas Sipopa and Lewanika's skills in retaining ultimate control over foreign trade and the circulation of charismatic goods—firearms included—that came with it, help explain why the kingdom did not experience the same decline as the Ruund and Luba states underwent, and why it would eventually negotiate its incorporation into emerging colonial structures at the end of the century from a position of comparative strength.

FURTHER CHALLENGES FROM THE SOUTH

A significant role in extending the sway of merchant capital over parts of the central savanna was also played by Portuguese settlements in Zambezia, in the lower Zambezi Valley in present-day central Mozambique. In the eighteenth century, perhaps the majority of the slaves obtained by the lower Zambezi's *prazeiros*—the Africanized descendants of the land-grant holders originally recognized by the Portuguese Crown in the seventeenth century—were retained on their estates (*prazos*) and put to productive uses.[75] Some of their number, however, were organized into standing armies and entrusted with the task of policing the sprawling concessions and enforcing the subjugation of both their free and unfree cultivators. Occupying a most ambiguous social location, armed slaves, known locally as *chikunda*, were both "the objects of domination and the means by which the *prazeiros* controlled the peasantry and accumulated wealth."[76] In time, the chikunda—who also worked for their owners as elephant hunters, slave raiders, and long-distance traders—gave birth to a distinctive ethnic identity, one structured around such cultural markers as a "disdain for agricultural labor"—which the Chikunda construed as the preserve of women, common slaves, and subjugated peasants—and the glorification of masculine, martial pursuits such as warfare and hunting, with the notable aid of imported and partly homemade firearms.[77]

Slave exports from Quelimane boomed early in the nineteenth century, when the port to the north of the mouth of the Zambezi began to attract Brazilian and other ships bent on escaping abolitionist efforts along the Atlantic coast of Africa.[78] The intensification of the southern slave trade meant that the Chikunda were no longer regarded as members of a comparatively privileged slave community, but, increasingly, as mere chattel for export. The Chikunda reacted violently to the new state of affairs, and slave insurrections and large-scale flights became the order of the day on the lower Zambezi. By the middle of the century, the prazo system had completely collapsed. Its fall resulted in the de facto emancipation of thousands of former armed slaves.[79] As Allen and Barbara Isaacman have expertly shown,[80] newly freed Chikunda had a number of options open to them in the new circumstances. Some continued to operate within the Portuguese sphere, working as professional hunters and porters for Luso-African traders (collectively known as *muzungus*). Other Chikunda—whom the Isaacmans refer to as "transfrontiersmen"—moved permanently away from the lower Zambezi. While some of these migrants ended up selling their military and hunting skills to vulnerable Chewa, Nsenga, and Gwembe Tonga communities, other Chikunda turned into state builders. Led by muzungu adventurers, Chikunda polities sprang up near the Luangwa-Zambezi confluence from the 1860s. Some at least of these new formations— especially those of the warlords Kanyemba and Matakenya—were veritable conquest states, resembling in many respects the creations of Msiri and other Congolese warlords. The products of the slave and ivory frontiers, Chikunda warlord states resorted to large-scale raiding and taxed local inhabitants mercilessly. With guns being deployed as their principal tools of commodity production, Chikunda conquerors made a significant contribution to regional instability, laying waste to large areas of present-day southern, central, and eastern Zambia.

Besides contending with the forces of merchant capital, the societies of the central savanna also had to deal with the extensive ripples of the South African *Mfecane*. Several reasons have been adduced to explain the turmoil that affected present-day KwaZulu-Natal and neighboring areas early in the nineteenth century. It has been argued, for instance, that conflict over scarce resources increased from about 1750 as a result of either demographic growth, climatic change, or both. Other scholars have preferred to relate competition among northern Nguni-speakers to the expansion of the trade flowing out of Portuguese-dominated Delagoa Bay in southern Mozambique.[81] Whatever its ultimate causes, increasing tension in KwaZulu-Natal precipitated processes of centralization and enlargement of political scale. Among the northern Nguni, state formation took one very specific course. Age sets, or *amabutho*, had been a distinctive feature of social organization in the area for decades, if not centuries. In their original form, the amabutho consisted of groupings of young men "brought together by chiefs for short periods to be taken through the rites of circumcision and perhaps to engage in certain services, such as hunting."[82] In the deteriorating political landscape of the late eighteenth century, local leaders transformed these age sets into labor and war regiments. So reconfigured, the amabutho took on the character of standing armies: they were both a consequence and a cause of the increasing level of militarism obtaining among the northern Nguni.

By the late eighteenth century, a number of opposing power blocs—the principal of which were the Ngwane, the Ndwandwe, and the Mthetwa—had emerged. Conflict between them eventually span out of control, precipitating the so-called Mfecane, a series of wars and migrations that transformed the sociopolitical landscape not only of KwaZulu-Natal, but of the broader southern African region as well. Among the protagonists of the turmoil were Sebitwane—the Sotho-speaking chief of the aforementioned Kololo, the migrant group who conquered and ruled the upper Zambezi floodplain and the Caprivi Strip between the early 1840s and the Luyana/Lozi *reconquista* of 1864—and Zwangendaba Jere, the war leader to whom the principal Ngoni groups of present-day eastern Zambia and Malawi trace their origin. Coming into their own in the second half of the nineteenth century, the conquering Ngoni kingdoms spawned by the Mfecane affected roughly the same areas as those into which Chikunda "transfrontiersmen" were then expanding. As will be seen in chapter 5, the Ngoni approach both to international commerce and to firearms, its most fundamental of by-products, differed from that of the Chikunda. But, even though the Ngoni's military preparedness had nothing to do with access to firearms, and was rather the result of the "meritocratic" aspects inherent in Ngoni age-set regiment systems, its effects undoubtedly magnified the violence and insecurity that accompanied the inland advance of the frontier of merchant capital.

↜

The primary aim of this chapter was to supply the reader with enough background data to engage with the chapters that follow. Given their imbrication in large-scale sociopolitical developments, firearms have already repeatedly cropped up in the discussion, providing some indication of the varied reactions they elicited, and the different meanings and functions attributed to them, across the central savanna. It is now time to explore these reactions and productions more systematically.

PART II

Guns and Society on the

Upper Zambezi and in Katanga

2 ⇜ The Domestication of the Musket on the Upper Zambezi

HOW DID the diversity-in-unity described in chapter 1 affect patterns of gun domestication in the interior of central Africa? This chapter begins to address the question by charting the relations between firearms and the peoples of the upper Zambezi, the area to which we first turn on account of its comparatively early exposure to the new technology. In the previous chapter, the point has already been adumbrated that central African responses to firearms were not uniform: different societies understood guns differently, attributing them culturally specific meanings and functions; the transformative effects of guns themselves varied accordingly. This chapter teases out some of the possible outcomes of cross-cultural technological consumption in the central savanna by contrasting gun holding among the rigidly hierarchical Luyana/Lozi of the upper Zambezi floodplain with the situation obtaining among the smaller-scale hunting and raiding groups to their north and northeast. In the second half of the nineteenth century, it will be argued, newly restored Lozi elites reasserted the sway of the monarchical dispensation following a period of foreign domination by centralizing the gun trade of the floodplain and by inserting firearms into royal symbolism. In more politically fragmented contexts, however, the domestication of externally introduced firearms conformed to a different pattern. The evidence presented in this chapter suggests that among Luvale and Kaonde hunters imported muskets were deployed less as means of political centralization and monarchical celebration than as individually owned tools of commodity production and markers of masculinity. Among the Kaonde, moreover, muskets were also reconstructed as a form of currency.

The chapter also tackles the difficult question of the practical effectiveness of the guns made available to central African societies over the course of the nineteenth century. In dealing with this "proverbial old chestnut,"[1] historians of such diverse territories as eighteenth-century Madagascar and nineteenth-century central Sudan have pointed to the technical deficiencies of imported guns in the context of arguments against overstating their overall significance.[2] Although such perspectives play down the historical centrality of firearms, their underpinnings, in keeping with much of the history of technology in Africa, remain fundamentally deterministic: if the technology was not sufficiently "developed," then its "impact" on society could not possibly have been profound.

The hold of these views is evident in the literature pertaining to the border region between present-day Zambia and Angola. Joseph Miller, for instance, regarded the "unreliability" of European trade guns—only a "small percentage" of which "survived the first few attempts to fire them"—as one of the principal causes of the continuity in military hardware and organization that characterized the Angolan interior in the eighteenth century.[3] In a similar vein, Achim von Oppen's study of the precolonial economy of the upper Zambezi and Kasai region in the nineteenth century presents the "remarkably poor" performance and "very limited durability" of the *lazarinas* or *lazarinos* (see figures 2.1 and 2.2), the often untested flintlock muzzle-loaders that dominated the trade between the central Angolan plateau and the Zambezi headwaters in the nineteenth century, as indications that neither the reported disappearance of elephants in the area from c. 1850 nor the depletion of game in general can be ascribed with any certainty to the spread of guns.[4]

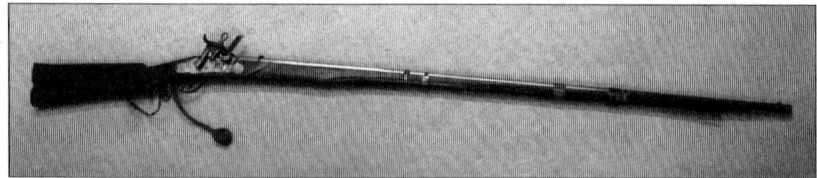

FIGURE 2.1 Lazarina, Musée Royal de l'Armée, Brussels, Belgium. Photograph by the author, July 2012. Initially produced in the eighteenth century by the Portuguese manufacturer Lázaro Lazarino Legítimo of Braga, by the middle of the following century, the time in which Liège began seriously to compete with Birmingham in the production of Africa-bound trade guns, the bulk of the muskets imported into Angola consisted of Belgian imitations. According to the explorer Serpa Pinto, the Belgian-made lazarinas employed and traded by the Ovimbundu of central Angola in the second half of the nineteenth century were "but a clumsy imitation of the perfect weapon turned out by the celebrated" Lázaro Lazarino. Alexandre A. da Rocha de Serpa Pinto, *How I Crossed Africa*, tr. Alfred Elwes (London: Sampson Low, Marston, Searle, & Rivington, 1881), 1:179. Perhaps because of its comparatively late production

date (Liège, c. 1900), the piece reproduced here would seem to belie Serpa Pinto's trenchant remarks. Unusually for African trade muskets, the barrel of this lazarina was tested at the Liège proof house, whose marks it bears, alongside the (fake) label "Lazaro Lazarino Legitimo Debraga." Obvious African inscriptions include the wooden stick used as a ramrod and the two amulets attached to the trigger piece. (See figure 2.2, below.) My thanks to Paul Dubrunfaut, keeper of firearms at the Musée Royal de l'Armée, for sharing information about the musket in his safekeeping. Courtesy of the Musée Royal de l'Armée, Brussels. See also Jean-Luc Vellut, "L'économie internationale des côtes de Guinée Inférieure au XIXe siècle," in *Actas de I reunião internacional de história de África: Relação Europa-África no 3° quartel do séc. XIX*, ed. Maria Emilia Madeira Santos (Lisbon: Instituto de Investigação Científica Tropical, 1989), 172–73; Emrys Chew, *Arming the Periphery: The Arms Trade in the Indian Ocean during the Age of Global Empire* (New York: Palgrave Macmillan, 2012), 86–90.

FIGURE 2.2 Lazarina, Musée Royal de l'Armée, Brussels, Belgium (detail). Photograph by the author, July 2012. One of the two charms is a string of dried snake skin; the other consists of an elongated pouch containing medicinal powder. Despite the patrimony of shooting skills accumulated by African gun users on the upper Zambezi, supernatural assistance was often considered necessary to ensure accurate firing and the success of the hunt. Courtesy of the Musée Royal de l'Armée, Brussels.

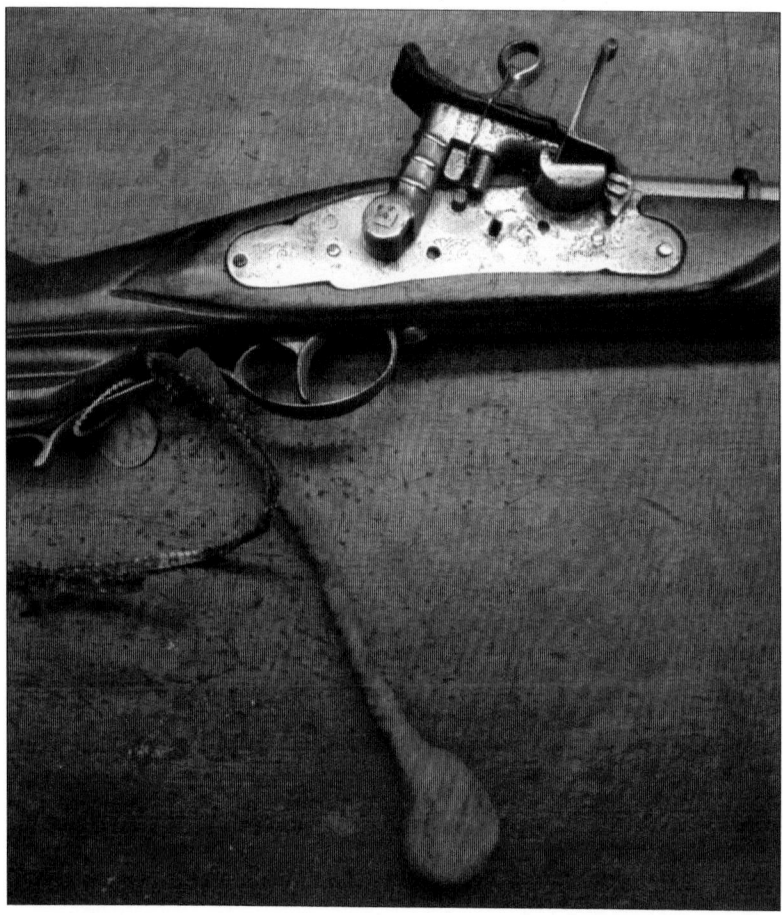

This chapter offers an alternative reading of the available evidence. By reexamining the movement and dynamics of the gun frontier in the upper Zambezi area, it contends that scholarly opinion that minimizes the centrality of imported guns on account of their technical shortcomings is misleading on at least three counts. First, the alleged weaknesses of the new technology are hard to reconcile with the upper Zambezi's unquenchable demand for European-made guns from the early nineteenth century onwards and the fact that, as Miller himself is aware, firearms always constituted "the 'very soul of commerce' in the exchange of people for goods" in the Angolan interior. "No Africans dealing with the foreigners," Miller elaborated, "could pass up the opportunity to acquire a gun and a supply of powder, and Africans would refuse all other wares and withdraw from negotiations if the assortment of goods offered lacked these essential components."[5] If firearms had really been invariably inefficient, and therefore of only marginal economic and military significance, then it is not at all clear why the bulk of the inhabitants of the region consistently insisted on obtaining them throughout the era of the long-distance trade.[6] To put it differently: it behooves historians to try to account for the reported African "passion" for muskets and for their "indispensability" to trade in the interior.[7]

Second, as already pointed out by Maria Emilia Madeira Santos, an exclusive emphasis on the technical shortcomings of trade guns trivializes the proficiency of nineteenth-century central African master ironsmiths.[8] Far from being the unchangeable "relic[s] of a very distant past," African iron-workers—Colleen Kriger has powerfully argued—displayed a clear and sustained ability to react to changed historical conditions, adapting their work and techniques "to new surroundings and opportunities. . . . Their history involves more than the initial introduction of iron into society, and ironworking was not a complete, self-contained package of tools and processes that was introduced once and then simply continued for centuries."[9] The sparse but compelling evidence examined in this and the next chapter suggests that African ironworkers not infrequently possessed the specialized skills and knowhow necessary to overcome or minimize the inherent deficiencies of the new technology.[10] These skills, no doubt, contributed to enhance such already considerable social prestige and professional autonomy as they had attained over the course of previous centuries. As remarked by Rory Pilossof, the full history has yet to be written of the "lively small-scale firearm repair and service industry" that developed in precolonial Africa.[11] The data presented here reinforce Pilossof's contention.

Third, and most important, to stress the poor quality of a given technology is often to ignore the latter's "interpretative flexibility"—the fact, that is, that "there is no one essential use that can be deduced from the artifact itself."[12] Over the course of the nineteenth century central Africa's gun societies adopted firearms for an array of practical and symbolic uses. Going beyond the military and economic spheres, such uses were sometimes at variance with those for which the weapons had originally been manufactured in their European settings. In other words, in assessing the historical significance of guns on the upper Zambezi, what Clapperton Mavhunga calls the specific "roles that Africans gave to firearms in contexts internal to their circumstances" need to be explored alongside more predictable patterns of gun deployment.[13]

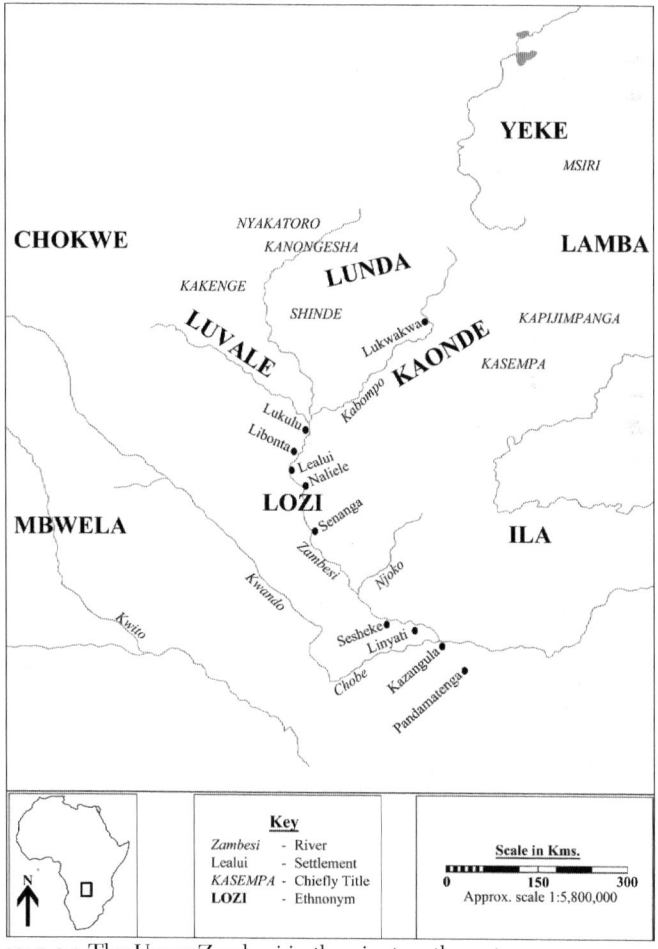

MAP 2.1 The Upper Zambezi in the nineteenth century.

By understanding guns in terms of their own varied idioms of power and sociocultural forms, the peoples of the upper Zambezi operated as agents of re-innovation. In the process, they created the conditions that enabled them to engage with a technology that, in other frameworks, might have been regarded as merely obsolete and inadequate.

SLOW BEGINNINGS IN BAROTSELAND

As has been argued in chapter 1, the appearance of the gun along the present-day border between Zambia and Angola was part of broader processes of violent socioeconomic change. The advancing frontier of the Angolan slave trade—centered on the Portuguese coastal town of Benguela and driven by Luso-African and Umbundu-speaking entrepreneurs (locally known as "Mambari")—reached the upper Zambezi area in the latter part of the eighteenth century. Further regional transformations owed less to the Atlantic economy than to the northern ripples of the demographic dislocations of the South African Mfecane. The most dramatic population movement to affect the region under discussion was that which led a group of Sotho-speaking migrants and refugees, the Kololo, to overrun what has been presented in chapter 1 as the area's most complex and stratified polity, the Luyana kingdom of the upper Zambezi floodplain and surrounding districts, in the early 1840s.

Unlike Walima Kalusa, who has recently described Barotseland and the Caprivi Strip under the Kololo as awash with guns,[14] I view Livingstone and Silva Porto's overall paucity of references to European-made weapons in the area, the "immense" number of elephants and other game near both Linyati and Sesheke and in the floodplain,[15] the weapons' abnormally high prices,[16] and—last but not least—the reported poor marksmanship of the Kololo as indications that firearms were still rare in Barotseland in the 1850s.[17] A few Kololo royals had some "wretched" guns which they "wretchedly used,"[18] and King Sekeletu's opponent, Mpepe, the governor of Naliele, was given "a small cannon"—or "a large blunderbuss to be mounted as a cannon"—by Silva Porto in 1853.[19] But stabbing and throwing spears and shields made of hides remained the dominant Kololo weapons throughout the decade—as is also borne out by the facility with which another pioneer trader, the South African James Chapman, conned Ponwane, "the headman of Linyati," and other Kololo grandees in 1853. Having been asked to "repair some guns," he took advantage of their "ignorance" of Western arms, "selected 5 of the easiest and repaired them for a tusk worth £15, at which rate [he] pocketed [his] pride."[20] It was this same "ignorance" on the part of the Kololo that apparently prompted the Batawana chief Letsholathebe, of present-day northern Botswana, to challenge Sekeletu's authority by appropriating some of the latter's ivory. Having recently acquired guns, Livingstone explained, Letsholathebe now considered himself "more than a match" for the less well-armed Kololo.[21]

To say that firearms were scarce in mid-nineteenth-century Barotseland is not to say that the Kololo resisted their introduction. The opposite, in fact, was true, for the conquerors of the Luyana state were clearly keen on European weapons, the destructive potential of which they had experienced at the hands of Griqua musketeers at the start of their northward migration in the 1820s, and which they expected to use to contain the threat posed by Mzilikazi's Ndebele, another product of the Mfecane, who had settled to the south of the Zambezi River.[22] It is clearly significant that when he first met Livingstone's party in 1851, Sebitwane, Sekeletu's father and predecessor, was convinced that "our teaching was chiefly the art of shooting . . . and that by our giving him guns he would thereby procure peace."[23] Four years later, "two superior double barrelled rifles" and "newly invented bullets for . . . shooting elephants" topped the list of foreign goods that Sekeletu wished to obtain from Livingstone upon his return from the Mozambican coast.[24] Thus, the fact that Barotseland remained lightly armed until at least the early 1860s ought not to be ascribed to cultural opposition to the new technology (as would be the case with the Ngoni of eastern Zambia and Malawi; see chapter 5). More simply, it was the result of the area's comparatively late incorporation into the Atlantic trading network. This, in turn, was a consequence of the very specific requirements of the upper Zambezi floodplain's political economy. The internal need for slaves for agricultural purposes and public works had led the last pre-Kololo Luyana king, Mulambwa Santulu, to shut his country to Mambari slave and gun traders in the early nineteenth century.[25] It was only after the death of Mulambwa, the civil war that followed it, and the related Kololo conquest in the 1840s that Mambari traders returned in force to Barotseland, whose new rulers were now ready to export some war captives and, more importantly, ivory to the Ovimbundu plateau and the Angolan shoreline. The ideal conditions faced by Silva Porto in early-1850s Bulozi, where ivory was cheap and plentiful and foreign wares scarce and expensive, prove that the long-distance trade was still in a relatively embryonic stage,[26] and that the Kololo, for all their eagerness to engage with foreign imports, had not yet bridged the technological gap that separated them from some of their immediate neighbors.

GUNS AND LUVALE MEN

In the early nineteenth century, then, the upper Zambezi floodplain had been all but closed to Angolan long-distance operators. They, however, had found more willing partners among the highly mobile hunting societies near the headwaters of the river. The ways in which these latter deployed and understood guns in the late precolonial era form the subject of this section. The first recorded trading visit to the Luena or Luvale ("Lovar"), then living mainly along the middle Luena River, took place in 1794–1795. At the time, the people of the *Kakenge* ("Caquinga") and the *Chinyama* ("Quinhama") were already said to be "warlike" (muito inclinados a Guerra), but still only armed with "bows and arrows, spears and knives, and wooden shields" (Arco e Frexa, Zahaia, e Facaõ, e Escudos que fazem de páo).[27] They had no firearms, a slightly later report elaborated, "because they [did] not know how to use them" (por não saberem uzar dellas).[28] The situation, however, evolved rapidly, with gun and gunpowder imports increasing in direct proportion to slave and, from the 1830s, ivory and beeswax exports. By the early 1850s, the Luvale, not unlike their western neighbors, the Chokwe, had accumulated "many guns,"[29] and they were among the most important suppliers of slaves to Angola.[30] The intensification of contacts with the Atlantic world left unmistakable imprints on the language of the Luvale: one of the numerous Portuguese-derived words they adopted was *mbalili*, from the Portuguese for "powder-keg," *barril*.[31] The then Kakenge boasted a long connection with Mambari traders, by whom, however, he was feared, for he and other "wild Luvale" (wilden Ka-lóbar-Völkern) were not averse to attacking passing caravans with their weapons.[32] It was probably this same Kakenge who is remembered in local traditions as a harsh and "fierce chief," famous "for guns and gunpowder," known as *fundanga*, another Angolan borrowing, to the Luvale. "He had many, many slaves and he bought and sold many people."[33]

Numbers, however, tell only part of the story. What really matters are the uses to which the Luvale put their abundant lazarinas and the ways in which these guns interacted with preexisting politico-economic and cultural structures. The relationship between hunting and the spread of firearms among the Luvale was complex and dialectical: if well-developed autochthonous hunting traditions (the Luvale had as many as ten words for different types of arrowheads[34]) facilitated the rapid adoption of guns, firearms themselves transformed and strengthened such traditions. As in mid-nineteenth-century Botswana, "the acquisition of guns was both a cause and a consequence of a surge" in the upper Zambezi's hunting trade.[35]

A critical consideration is that smoothbore muzzle-loaders, in general, and the lazarinas, in particular, constituted an accessible technology.[36] Since they were typically made from "soft" wrought iron (as opposed to "hard" steel), preexisting ironworking skills could be so refined as to greatly prolong the lifespan of a damaged weapon and/or keep a defective one in serviceable order. Livingstone, for one, thought that some of the locks of the lazarinas used by the Luvale could be as old as "100 years."[37] The proficiency of upper Zambezian gun-menders is exemplified very clearly by the Luvale's immediate neighbors, the Chokwe, one of whose blacksmiths was once put to the test by the Hungarian trader and explorer László (Ladislaus) Magyar by being given "a damaged musket without lock for repair, and as a sample a French flintlock on whose cover the word 'Laport' was engraved. After some days the blacksmith returned the musket; he had not only manufactured the lock well and neatly, but also faithfully engraved the word 'Laport' on the cover, with the letters being only somewhat less subtle."[38] Still in the 1850s, Silva Porto was much impressed by the technological knowhow of the Mbwela, another neighboring group. They, wrote the famous *sertanejo*, "handled firearms tolerably well. With the exception of the barrel, they manufactured them perfectly, better than any other tribe."[39]

Even though barrels could not be fabricated *ex novo*, they could certainly be mended.[40] And while gunflints (*matale*, in chiLuvale) were produced locally, because of their heavy caliber, single-shot smoothbore muzzle-loaders could also be loaded with most kinds of homemade projectiles—including "bullets of common iron about the size of a pea"[41]—and easily used in association with earlier hunting equipment.[42] It is certainly not coincidental that the general word for "gun" in chiLuvale—*uta*—is the same as that for "bow." Perceived structural similarities went further: one of the four Luvale words for "gun trigger" was *lukusa*, also used for "bowstring," while *kunana* meant both "to draw bow" and "to take aim with a gun."[43] As construed by the Luvale, then, flintlocks—for which they had at least four different words[44]—belonged to the same category as preexisting hunting tools, the continuing use of which the lazarinas did not preclude. When all these factors are borne in mind—and due allowance is made for the patrimony of shooting skills that the Luvale, their common reliance on "gun medicine" notwithstanding, are likely to have accumulated through sheer practice[45]—it is easier to understand why the lazarinas became a crucial tool of production around the headwaters of the Zambezi, one whose extensive deployment would have significant environmental consequences.

To be sure, when describing the situation obtaining among the southern Lunda of the *Shinde*, the Luvale's antagonists, Livingstone asserted that "their bows and arrows [had] been nearly as efficacious in clearing the country of game as firearms."[46] Also, he quickly became aware of the extent to which prolonged damp conditions could affect the performance of muzzle-loaders.[47] Yet he remained convinced that the lazarinas did give Luvale hunters a distinct advantage over the less well-armed Lunda. In comparison with the Luvale, who "enjoy[ed] the privilege of hunting on both sides" of the upper Zambezi, the "Balonda [were] able to do little." This remark dates from July 1855, when he also noted that the Lunda-controlled left bank was still richer in game than the right one, which had been "much hunted by the Balobale who have guns."[48] Without being dismissive of traditional elephant-hunting techniques, the skills demanded by which he had had the chance to admire in Linyati,[49] Livingstone was also in no doubt as to the ultimate consequences for elephant herds of the diffusion of firearms. The Kololo had only just begun to hunt elephants with guns, he wrote in 1853; if they continued, "very soon none will appear in this part of the country. They retire before the gun sooner than any other animal."[50] Indeed, environmental degeneration was rapid among the gun-rich Luvale, for it was at about this time that the middle Luena seems to have exhausted its ivory supplies. Thereafter, whatever little ivory the Luvale continued to export alongside slaves and beeswax came mainly from newly conquered southern Lunda territory.[51]

Hunting and warfare have been described as "intimately connected" activities, not least because "the hunt is often the training ground for war" and "shooting skills that developed in one setting could be transferred to the other."[52] Livingstone might well have subscribed to this view, for, *contra* much scholarly literature, he never questioned the military significance of firearms, going so far as theorizing that, by making local conflicts "more terrible"—and by reducing the gap between "the strong and the brave," on the one hand, and "the weak & cautious," on the other—guns would work towards reducing the incidence of war in central Africa.[53] The available written sources do not permit us to gauge the manner in which the lazarinas were incorporated into Luvale tactics and organization. On the basis of oral evidence alone, Robert Papstein concluded that warfare on the upper Zambezi remained the affair of small bands of marauders bent on taking the enemy by surprise. But even if one accepts this hypothesis (and the unsubstantiated contention that the gun was more often used in war "as a blunderbuss rather than as a precision instrument"[54]), there is no doubt that the wide availability of firearms increased Luvale military potential. This is borne out by their reported ability to repel a Kololo party in the 1840s,[55] and, especially, by their successes during the nineteenth-century phase of the so-called Wars of Ulamba, in the course of which gun-wielding Luvale slave raiders attacked—and gradually encroached upon—the territories of the southern Lunda of the *Shinde*, to the east of the Luena, and of the *Katema*, the *Kapenda*, and others, to the northeast.[56]

By the time of Livingstone's passage in the 1850s, the Lunda of the *Shinde* had been in touch with Mambari traders for almost as long as their Luvale enemies.[57] Yet, for reasons that remain unclear, they had been less successful in modernizing their armament. Livingstone saw hardly any guns among the then Shinde's southern subjects, between the site of present-day Zambezi (colonial Balovale) and the Lufige River.[58] The "Portuguese guns" that he did spot in the Lunda capital on the Lufige, just inside present-day Angolan territory, were so poorly handled by their untrained owners that they constituted "less formidable weapons than the bows and arrows of others."[59] Given this seeming lack of competence—itself probably an indication of the weapons' comparatively recent introduction in the area—the Lundas' best responses to Luvale aggression were to beef up their ranks by absorbing fleeing Luvale refugees and to strengthen the defenses around their villages and even individual households.[60]

By c. 1870, the lazarinas had fully permeated Luvale society, changing it and being changed by it. In 1875, the small party of armed Luvale hunters whom Cameron met to the north of Nana Kandundu, the *Nyakatoros*' new capital in former Lunda territory, were very intrigued by the explorer's breech-loading rifle. But even though they examined it "with much admiration," "they did not consider it sufficiently long, their own weapons being lengthy Portuguese flint-locks" (see figure 2.3). It took a practical demonstration of the rifle's penetrative power and accuracy to dispel the hunters' knowing skepticism[61] — one that distinguished them sharply from the gun-poor Kololo, who, as we have seen, placed a blind trust in Chapman and other foreign "experts" in the new technology.

The Luvale prowess in hunting and warfare, in sum, was certainly driven by their familiarity with imported muskets. But the significance of the lazarinas was no less symbolic than material, for the new

FIGURE 2.3 Firearms, Livingstone Museum, Livingstone, Zambia. Photograph by the author, January 2012. Note the length of the lazarina (*first from right*) vis-à-vis other types of guns. The explorer Cameron may not have understood that the Luvale hunters whom he encountered near Nana Kandundu's had every reason to value the length of the barrels of their flintlocks. The trade gunpowder (*fundanga*) that they used needed more time to burn (and therefore more room in the barrel) than high-quality powder. Had the barrel been shorter, half of the gunpowder would have spurted out of the barrel unused when the gun was fired. The lazarina in the photograph — mistakenly described as being "of Italian manufacture" by G. Tylden — misses both the upper jaw of the lock and the frizzen. G. Tylden, "The Gun Trade in Central and Southern Africa," *Northern Rhodesia Journal* 2, no. 1 (1953), 47n3. The brass barrel bands are local repairs. According to Paul Dubrunfaut (personal communication to author, 30 June 2012), the heavy barrel in the (Liège-made?) specimen under discussion suggests an early nineteenth-century fabrication date. Courtesy of the Livingstone Museum, Livingstone.

technology also affected the realms of ideas, quickly becoming implicated in issues of gender identity. The relationship between gun ownership and masculinity among the Luvale was much in evidence in 1895, at the time of François Coillard's visit to the then Kakenge, who lived very close to the spot where Livingstone's Shinde had resided some forty years earlier. On the day that preceded his first meeting with the chief, the missionary and his Lozi escort were made the target of repeated hostile demonstrations on the part of "young men all armed with guns." After a tense night punctuated by war dances and continuous discharges of firearms, Coillard was finally brought to the presence of the chief. "The place was full of men, decked in their war-paint, and surrounded by bundles of guns." The menacing atmosphere eased a little on the following day, but the Kakenge still warned Coillard not to "take the Balubale for women," and he then proceeded to shock the missionary with the following example of "corporeal discourse":

> Suddenly, he threw himself backwards, stiffened himself, kicked about, tore himself with his nails, and made frightful contortions, rolled his eyes, ground his teeth, and uttered horrible cries. Then, calming himself with equal suddenness, he rose, and darted into his court. We remained stupefied! I had thought at first that the man had a fit, and I wanted to fetch some water; but all his people had risen, and were uttering savage cries to applaud him. Then I realised that Kakenge was vaunting his courage by imitating a ferocious beast struggling with his prey and devouring it.[62]

The link between guns and masculinity among Luvale hunters and raiders was confirmed a few years after Coillard's passage when Colin Harding explicitly reported that it was "unusual and so rare an occurrence to see a man without a gun [at the Kakenge's], as it [was] to encounter a woman without her infant."[63] The erection of a Portuguese fort — "well constructed, and garrisoned by a force of about 27 native soldiers, armed with Sniders, and possessing several spare M[artini] H[enry] rifles" — had done nothing to change the bellicose behavior of the town's male inhabitants. To the dismay of the same visiting British official, the latter, "armed with their Portuguese flint lock guns" and "girded with a miniature magazine in the shape of a small black bag containing spare flints, powder and caps," were wont daily to "parade the camp, assuming an insolent demeanour of the most intolerable character."[64]

Because of the reasons discussed above, firearms made a comparatively late entry into Barotseland. Once the domestication process began, however, it proceeded swiftly. Its shape and direction, this section will argue, were clearly influenced by the special characters of Lozi society and politics. Following the overthrow of the Kololo in 1864, Sipopa, the restored Lozi monarch, made contact with ivory traders and hunters from the west and the south.[65] His aim was to embark on a program of accelerated military overhaul with a view to consolidating his still fragile internal position and reasserting the tarnished grandeur of the old floodplain kingdom. In December 1865, the weapons available to Sipopa's soldiers still consisted only of "knobkerries, axes, and assegais" (cacheiras, machadinhas, e azagaias).[66] Less than two years later, however, a military *exercicio* overseen by the king already involved repeated discharges of musketry (though Silva Porto thought there was "no method in this way of training in the art of destruction").[67] By the mid-1870s, Barotseland's firepower had increased exponentially. Having spent several months in Sesheke in 1875, the Czech explorer Emil Holub hazarded an estimate of "the number of guns that had been introduced into the country from the south and west" over the course of the previous few years. This, he believed, "amounted to 500 flint muskets, 1,500 ordinary percussion muskets, eighty percussion elephant-guns, 150 rifles, thirty double-barrelled guns of various sorts, ten breech-loaders, and three revolvers."[68]

By the time of Holub's visit, the Lozi, unlike the Kololo, were routinely hunting elephant with guns, though they still used them in tandem with locally produced iron "elephant assegai[s]."[69] Besides dancing and drumming, the royally sponsored ceremonies that preceded elephant hunts now involved much firing in the air. On at least one occasion, the hunters employed by Sipopa also entertained the king by lifting a small pot of cooking elephant meat with the barrels of their guns.[70] During the great hunt of 1875, as many as "10,000 shots" were said to have been fired in the general melee that followed Sipopa's ill-timed first shot.[71] Before long the fauna of Barotseland—just like that of the Luvale heartland a few decades earlier—began to show signs of exhaustion.[72] By the mid-1880s, Lubosi Lewanika, the Lozi king since 1878, was clearly concerned by the environmental consequences of the widespread adoption of firearms in hunting. It was probably for this reason that, in 1886, he forbade the use of guns during the annual royal hunt.[73]

Lozi rulers drew on a culture of political and economic centralization that set them apart from the more fragmented Luvale chieftainships to the north. It is therefore not surprising that the former exerted a tighter control than the latter over external trade and the movement of firearms that came in its wake. Not only did Sipopa resolve to spend most of his time in Sesheke, to the south of the floodplain proper, with a view to being closer to Westbeech and other traders from the south,[74] but his monopolistic ambitions also led him to forbid most kinds of exchanges between his subjects and visiting merchants.[75] Ivory, which Sipopa accumulated through the payment of tribute or by sponsoring the activities of hunters, was considered "crown-property, and it [was] a capital offence for anyone to carry on any transactions with regards to [it] on his own account."[76] The trade in guns was another exclusive royal prerogative.[77] Distribution, too, was closely monitored. Unlike other products—Holub was told—firearms were never permanently given away by Sipopa, but only "lent" to hunters, chiefs, and subjects with the proviso that they could "be recalled at any moment at the royal pleasure."[78] Holub himself saw Sipopa personally "distributing powder and shot" in his "royal courtyard" on the eve of the hunt of 1875. The place "was so full of men equipped for the excursion that [he] could only with difficulty make [his] way across."[79]

So prominent were firearms in the entourage of Sipopa—who also used them, alongside "Marutse weapons, elephants' tusks, and various articles of apparel," to decorate the walls of one of his residences[80]—that Holub was led to believe that the second most important "officer of state" in the Lozi kingdom was Masangu, the "governor of the arsenal," whose principal responsibility was the "supervision of the ammunition and guns distributed to the vassals." "He was likewise superintendent of all the native smiths. I found him employed in repairing a gun, for which he was using hammers, chisels, pincers, and bellows, all of his own making, and of the most perfect construction that I had yet seen in South Africa."[81] Masangu had seemingly been trained by one of Silva Porto's own Ovimbundu gun-menders. By the early 1880s, this highly specialized Lozi artisan was able to carry out "any repairs of firearms on behalf of his king" (qualquer conserto de arma por ordem do suzerano).[82] This competence, of course, was a recent development in Barotseland, and it attests powerfully, not only to the increasingly significant position of firearms in the region's political and economic life, but also to the effectiveness with which the people of the upper Zambezi floodplain were learning to overcome some of the new technology's limitations by honing preexisting ironworking skills. There is, indeed, some evidence that Sipopa was closely associated with firearms in the minds of his people: when the king was ousted in 1876, one of the rebels' first actions was to throw "the great bulk" of his guns into the Zambezi.[83] Claiming to be invulnerable to bullets, Sipopa, the first gun-rich Lozi monarch, was eventually killed by a bullet through his chest.[84]

Lubosi Lewanika, who wrested the kingship from Mwanawina, Sipopa's immediate successor, in 1878, started off with comparatively few guns and powder, as evidenced by his attempt to requisition all the arms in the possession of Serpa Pinto, who visited the royal capital of Lealui in August–September of the same year.[85] When the Portuguese explorer's camp was set on fire during an altogether unclear episode, the Lozi attackers were apparently only armed with shields and "murderous assegais," and Serpa Pinto's men were able to keep them at bay by means of the "sustained fire" of their Snider-Enfield breech-loading rifles and an elephant gun charged with explosive bullets.[86] However, by the time he raided the cattle-rich Mashukulumbwe or Ila in 1882, Lewanika had certainly obtained some guns, for the exploits of the "weapon with the lightning" made an obvious impression on its victims.[87] His arsenal must have increased thereafter, partly in direct response to continuing internal opposition to his rule. The Jesuits who sought unsuccessfully to gain a foothold in Barotseland in the early 1880s were left in no doubt as to where Lewanika's priorities lay: gunpowder and guns topped the list of foreign goods that the king expected the missionaries to be able to manufacture.[88] Despite his earlier association with Sipopa, Westbeech traded with Lewanika both before and after the king's temporary removal from the throne in 1884–1885. After a long absence, Silva Porto, too, made his way back to Bulozi between the end of 1883 and the beginning of 1884. In exchange for Lewanika's ivory, he "brought with him a large quantity of calico, guns and powder."[89] Lewanika's trading partners, of course, also included independent Ovimbundu, some of whom are said to have brought their guns to bear in the decisive battle between the ousted king's partisans and his opponents late in 1885.[90] It was perhaps these same traders-*cum*-mercenaries who built the king an impressive residence in Lealui following his restoration.[91]

Lewanika's trading policies mirrored Sipopa's. In June 1886, Holub, on his second trip to the north of the Zambesi, became acquainted with

Liomba, the "trade minister" whom Lewanika had sent to intercept Westbeech at the Kazungula ferry and make "all the purchases for the royal household as well as to purchase all arms for the empire on the king's account."[92] A few months later, Watson, one of Westbeech's partners, bought some ivory from the headman of Sesheke. Both parties were aware of committing "a capital crime . . . since all ivory is private property of the king, which he uses, in turn, to buy arms and ammunition for all his subjects."[93] And firearms Lewanika did continue to buy throughout the late 1880s, for Coillard was "astonished" by the number of guns available to the Lozi at the time of their second major raid against the gun-poor Ila early in 1888. Even though he thought the assegai was "still the national weapon," the missionary could not help noticing that the Lozi now had arms of "every calibre. To be sure, they are not the most modern pattern; the majority are flint-locks. Never mind, they are *guns!* And to a Morotsi the name alone is magic."[94] Lewanika's enduring interest in guns—which the Lozi were now undoubtedly employing in warfare, alongside hunting—is also attested by Selous, who in September 1888 offered the king "a very good hammerless shot gun,"[95] and by the nature of the presents given to him by concession-seekers Ware and Lochner in 1889 and 1890, respectively. Invariably, these included considerable quantities of Martini-Henry rifles and ammunition.[96] It was in modern weapons that Lewanika planned to invest the annual payment of £2,000 that Lochner promised the British South Africa Company would make available to him in exchange for the concession of mineral rights within his possessions.[97] Late in 1893, when the Lozi, fearing a Ndebele invasion from the south, mobilized the army, European missionaries were once more "shocked by the quantity of guns present in the country. There were hundreds of them, mainly flintlocks and pseudo-muskets imported by the Mambari."[98] In fact, so well armed had the Lozi become that, about two years earlier, they had felt strong enough to take on the Luvale, some of whose leaders they brought under their sway after an expedition prompted by the then Shinde's request for help.[99]

Barotseland's centralized political system and monarchical tradition underlay local understandings of firearms. As pointed out above, the Lozi kings' manifest monopolistic tendencies found no equivalent among Luvale trading principalities. On the headwaters of the Zambezi, competing Luvale leaders were never able to prevent ordinary villagers from participating in the market economy, as epitomized by

Mambari caravans. As a result, the trade's by-products, including firearms, spread well beyond chiefly strata.[100] In Barotseland, on the other hand, access to firearms, powder, and ammunition was always closely dependent on the *Litungas'* patronage. Because of this, gun use among the Lozi was more elitist than in Luvale country, where the lazarinas represented the common man's weapon of choice and a critical component of his social identity. Among the Lozi, conversely, firearms were construed less as an attribute of masculinity than as a means of political centralization and a symbol of royal wealth and proximity to the royal court. Linguistic evidence might also point in the direction of elite control over firearms, since *tobolo*, one of the words for "gun" in Silozi, the mixture of Sesuto and Siluyana spoken by the Lozi from the last decades of the nineteenth century, was also the word for the dowry that elders were expected to make available to their juniors.[101]

The evidence surveyed above suggests that both Sipopa and Lewanika regarded firearms as instruments of rule, not only because of their potential for ensuring physical control, but also on account of their symbolic attributes. By centralizing the gun trade into their hands and fitting firearms into their court ceremonial, the two Lozi kings—not unlike Elizabeth I of England some three centuries earlier—must have aimed at exploiting the "expressive hegemonic power" of artifacts.[102] Firearms and other imported objects—such as the western apparel that the two monarchs commonly donned[103]—could be made to communicate their "legitimacy of rule, aspirations for the kingdom, qualities of power and majesty, and, finally, godlike status."[104] In conveying a vision of state-controlled technological prowess, moreover, firearms served to reassert the modernity and worthiness of a social order that political turmoil and foreign invasion had recently called into question. Herein lay the real "magic" of firearms among the Lozi.

A KAONDE CURRENCY

Of the small-scale societies to the east and northeast of the upper Zambezi, on the furthest reaches of Mambari commercial penetration, the cattle-keeping Ila and Lenje remained lightly armed throughout most of the nineteenth century. As a result, they both suffered from joint Kololo-Ovimbundu raids in the 1850s,[105] while the Ila were also targeted by Lewanika's Lozi in the 1880s (see above). Other groups, most notably the Kaonde of present-day Solwezi and Kasempa districts, in northwestern Zambia, responded more readily to the advent of gun technology. As in the case of the Luvale, the ease with which the Kaonde integrated firearms into their cultural settings had certainly much to do with the strength of their preexisting hunting traditions. The evidence relating to the Kaonde, however, is especially interesting since it illuminates yet another central African understanding of firearms.

The traditions or reminiscences recorded by Frank Melland in the early 1920s suggest that some guns were being used in warfare by the Kaonde of the *Kapijimpanga* and the neighboring Lamba of the *Mulonga* and the *Kalilele* in the latter part of the nineteenth century.[106] Near the modern Congolese border, the aggressiveness of Msiri's Yeke (see chapter 3)—still vividly remembered in the late 1960s—must have also contributed to accelerating the process of Kaonde rearmament.[107] More details are available for southern Kaonde groups. Around present-day Kasempa, firearms were relatively uncommon until the late 1870s. When he killed his cousin Kabambala to assume the dignity of *Kasempa*, Jipumpu is said to have had "only one muzzle loader" at his disposal.[108] However, thanks to both his temporary alliance with Yeke raiders and partnership with Mambari traders, Jipumpu—who is also remembered as a great elephant hunter—must have been able to beef up his arsenal quickly and effectively. Between the 1880s and the 1890s he became an assertive warlord in his own right, raiding some Ila communities for slaves, expelling the Nkoya leader Mwene Kahare from his capital, and even defeating an expedition sent by Lewanika c. 1897 to support Jipumpu's rival, his fellow Kaonde chief Mubambe Mushima.[109] Jipumpu's victory owed something both to the fact that most of his men were now "armed with muzzle loaders" and to the impregnable nature of his stronghold on Kamusongolwa's hill, a high "kopje . . . well provided with good deep caves where most of the people could entirely hide themselves from being shot."[110] In fact, by this time, "all the villages" around Kasempa's were surrounded by stockades "of considerable height and strength." Some of these fortifications had "manholes or dug-outs . . . on the inside of the breastwork every dozen yards or so from which the defenders would emerge and discharge their guns, retiring again into the pits to reload."[111]

Like chiLuvale, kiKaonde illustrates the intimate relationship that existed between hunting and firearms in the minds of its speakers. As in the former language, the general word for "gun" in kiKaonde—*buta*—is the same as that for "bow," while *mukanda* denotes both a "bowstring" and the "trigger of a gun." Moreover, one at least of the three Kaonde words for bad or unsuccessful hunter, *kayenge*, was synonymous with "poor shot."[112] But from the late nineteenth century, the Kaonde—among whom cattle, the key form of transferable wealth in the territories to the south and southwest, could not thrive on account of the presence of the tsetse fly—endowed their muskets with a less easily foreseeable function, beginning to conceive of them "as a form of currency . . . serving most of the purposes . . . for which other natives use[d] cattle or slaves."[113] Gradually, "all transactions regarding wives, inheritance, succession, compensation, illness, deaths, burials, and initiation ceremonies" came to entail "the loaning or passing of guns and powder."[114] In using muskets not only as sureties (*kapopoka-milonga*[115]), but also as standards of value for social payments and even everyday transactions, the Kaonde were either inaugurating an entirely novel system or, perhaps more probably, modernizing an earlier tradition of regulated exchanges of hunting weapons and/or metal currencies. It is also possible that the use of muskets as currency attenuated the technology's gendered character: at least in the early 1920s, female gun owners were not unknown.[116] What is certain is that such practice explains the extraordinary extent to which the muzzle-loader embedded itself in Kaonde social structure and culture.

This profound technological engagement also had religious reverberations, as Kaonde hunters—accustomed, like most other savanna hunters were, to propitiate the ancestors both on the eve of and during the hunt—now began routinely to use their guns to invoke the same spirits whose blessing was being sought. This, *inter alia*, accounts for the dilemma faced by an early Christian convert, Kimengwa, who, having been preliminarily appointed to a headship near Kapijimpanga's, was requested not only to inherit the widow of his predecessors, but also to "place his gun at the crossroads" and "to invoke the dead man whom he was to succeed, asking him to give the gun power to kill [the] animals" whose meat would then feature in the succession ceremony. Kimengwa, an admiring missionary reported, refused to comply, "saying that if he could kill in his own strength he would, but he would have nothing to do with invoking the dead."[117]

A paradox underlies this chapter, for the adoption of a nondeterministic perspective results in the attribution of greater significance to firearms than has commonly been the case in the relevant specialist literature.

Guns—this chapter has shown—spread throughout the border area between Zambia and Angola in the course of the nineteenth century. The enthusiasm with which the imported technology was taken up by most of the peoples of the region was in large measure the result of their ability successfully to deploy it for a variety of both innovative and predictable purposes. In the gun societies examined in this chapter, firearms acquired different meanings, and the modalities of their domestication were invariably informed by local circumstances and power relations. Among the monarchical Lozi after the Kololo interlude, guns were primarily understood as symbols of royal power and means of political centralization. Among Luvale hunters and raiders, muskets became defining markers of masculinity, while in Kaonde country they were also recast as a polyvalent form of currency. However, throughout the region under discussion, imported muskets hardly lost their immediate service functions, and were consistently deployed as hunting and military tools. This was because, besides attributing culturally specific roles to firearms, gun users on the upper Zambezi also proved able to take advantage of the new technology's accessible nature and thus minimize its inherent shortcomings by honing preexisting metalworking competences. As chapter 4 will show, the initial weakness of the British administration in colonial North-Western Rhodesia meant that this heterogeneous process of technological engagement continued during the early years of the twentieth century, being only brought to an end from the early 1920s, when the hitherto unregulated right to possess and exchange guns was taken away from the peoples of the region.

3 ✎ The Warlord's Muskets

The Political Economy of Garenganze

IN THE second half of the nineteenth century, the Lozi elites of the upper Zambezi floodplain adapted to the requirements of international trade, harnessing its most significant of by-products—firearms—to promote centralization and drive forward a restored royalist project. While attesting to the resilience of central Africa's political traditions, the Lozi trajectory was, if not unique, then certainly distinctive. As has been argued in chapter 1, other ruling groups in the central savanna proved unequal to the challenges of the new era. The disappearance of Luba and Ruund/Lunda elites, or the fragmentation and contraction of their spheres of influence, opened the door to the emergence of warlord polities—predatory formations from which the central African interior had hitherto been sheltered. Often led by self-made, charismatic opportunists, these new political organizations were characterized above all by their openness to the market, military preparedness, and the readiness to resort to forms of extreme compulsion to obtain the commodities that fed the long-distance trade. Garenganze, the conquest state inaugurated by Msiri's Yeke (from *bayeke*, "elephant hunters" in kiSumbwa) in southern Katanga early in the second half of the nineteenth century, constitutes a relatively well-documented example of the spread of warlordism in our region in the age of merchant capital.

The origins and qualities of the Yeke state were reflected in its intense engagement with firearms. As the first two sections of this chapter demonstrate, the entire political economy of Garenganze was predicated on the availability of imported guns, which Msiri and his inner circle understood primarily as state-controlled means of military domination and commodity production. To be sure, as we know, even long-established monarchical elites had every interest in monopolizing access to imported firearms. But what was an ambition and a prop to power for the latter was a categorical necessity for upstart state-builders shorn of traditional support bases. While it is at least possible to imagine a gun-poor, but historically legitimate, Lozi king emerging unscathed from the perilous waters of the nineteenth century, the career of Msiri and his system of rule would have been scarcely conceivable before the advent of gun technology and the kind of politics that it enabled. By illustrating the inverse relationship that obtained between reliance on firepower and local sociopolitical rootedness, the trajectory of Garenganze substantiates this book's more general argument about the contingent nature of processes of technological transfer and domestication. Firearms were not predetermined to become the key to military and economic domination in southern Katanga in the second half of the nineteenth century; rather, the unquestionably fundamental role that they came to play was a result of the specific context in which they were deployed.

The limited value of deterministic understandings of the relationship between technology and society is further borne out by the fact that, for all their immediate impact, the many flintlock and percussion-lock muskets monopolized by Msiri proved insufficient to ensure the long-term stability of his political creation. The functioning of the driving sectors of Garenganze's political economy rested on the uninterrupted deployment of an imported technology. Though profitable in the short term, this dependence on outside sources of supply made Msiri's warlord state vulnerable to changes in external trading circumstances. The Yeke's predicament was compounded by their inability or, more probably, unwillingness to invest time and resources in the development of such techniques for the manufacture of gunpowder as were available elsewhere in the central savanna. By exposing these structural weaknesses, the Sanga rebellion of 1891—the subject of the final section of the chapter—brought Garenganze to its knees, paved the way for the European conquest of the region, and, as it were, reasserted the rights of society over technology.

THE EARLY HISTORY OF THE YEKE POLITY

Between the early 1850s and the late 1860s, Msiri, a Sumbwa caravan leader turned state-builder, succeeded in imposing his and his followers' sway over the territory between the Dikulwe and Lufira Rivers—a district that west-central Tanzanian caravaneers, including Msiri's own father, Kalasa, had begun occasionally to visit from the 1830s.[1] Before becoming the heartland of Garenganze, this copper-producing region—whose principal autochthonous titled leaders were the *Mpande*, of the Sanga, and the *Katanga*, of the Lemba—had formed the western

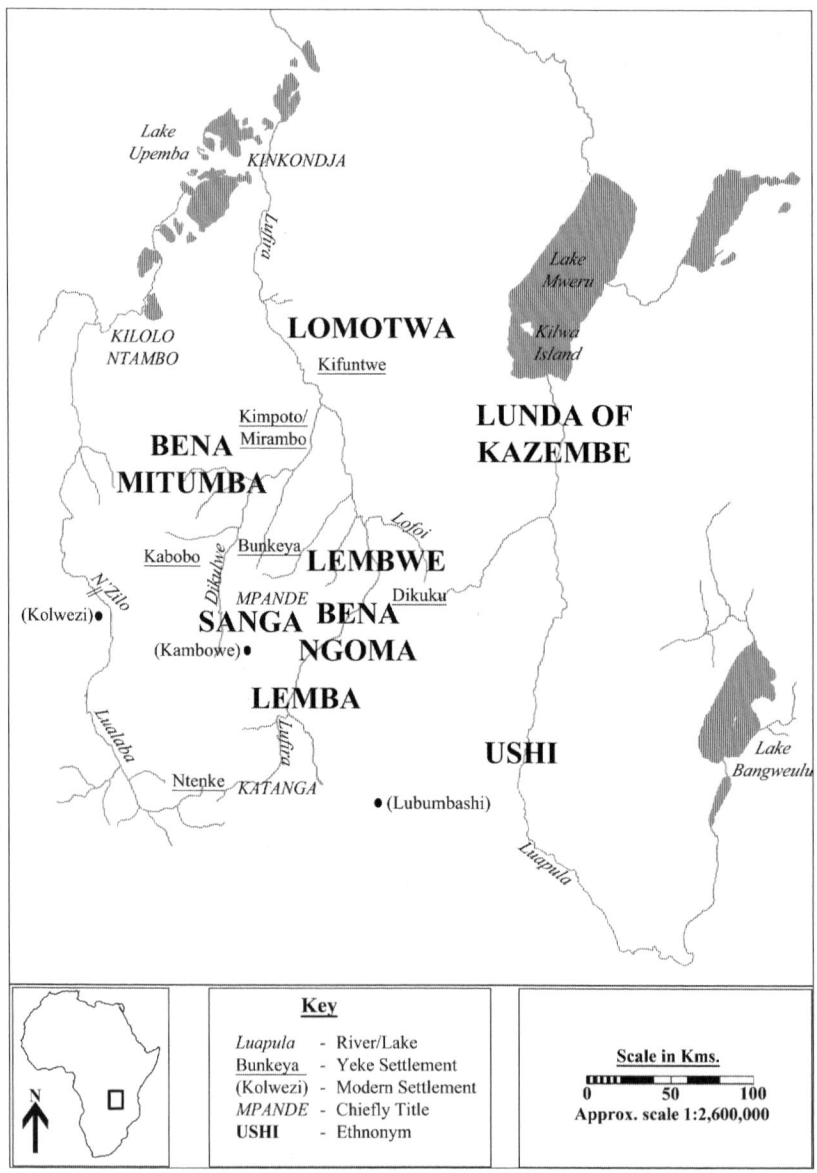

MAP 3.1 Southern Katanga in the late nineteenth century.

periphery of the eastern Lunda kingdom of Kazembe, founded one century earlier in the lower Luapula Valley by a conquering elite hailing from the southeastern periphery of the Ruund state (see chapter 1).

Bent both on glorifying and exculpating the figure of Msiri, Yeke sources depict his early territorial conquests in the following terms. After his arrival in southern Katanga in the 1850s, Msiri carried out a series of military forays on behalf of preexisting local authorities, whose prerogatives he unfailingly respected and to whom he always deferred in such matters as the apportionment of war prisoners. It was with a view to rewarding Msiri for the services that he had rendered him that the then Mpande, Kya Mulemba, willfully handed him over his ancestral territory. On the day of the pivotal encounter, Mpande is reported to have greeted Msiri thus: "'I am happy when I see you, my child. Look at the good you did. This was once a troubled country. Now the country is yours! Give me your hand!' They shook hands. Mpanda [sic] traced a line on the ground between them and told Msiri: 'You too draw a line! . . . I give this country to you, my child. I love you!'" Lemba country fell into Msiri's hands through less peaceful means, but the Lemba were the architects of their own downfall. Unjustly accused of having caused the death of the then Katanga, Msiri was eventually forced to take offensive action. The Lemba were scattered, and their erstwhile territory was added to Msiri's recently acquired Sanga possessions.[2]

What role did firearms play in all of this? While registering Msiri's early successes, Yeke narratives do not explicitly dwell on his initial military superiority over his future subjects. Given the selectivity of the traditions of high political office (see introduction), this is scarcely surprising. Yet, even Mukanda Bantu—Msiri's successor from the end of 1891 and the source of a comprehensive account of Yeke history shortly before his death in July 1910—cannot entirely prevent guns from making themselves heard. Mukanda Bantu makes no mention of firearms in his depiction of Msiri's expeditions against local enemies of the then Katanga and Mpande, but he does hint at their importance in a subsequent clash with Mwata Kazembe VII Muonga Sunkutu (1862–1872), who in the mid-1860s dispatched a military contingent against Msiri in a belated attempt to preserve the integrity of the eastern Lunda tributary sphere. Lubabila ("Luvavila"), the leader of the Lunda army, is remembered as having been killed by a gunshot personally fired by Msiri, who then celebrated the setback of his principal regional antagonists by taking over the Sumbwa royal title of *Mwami* and setting about organizing the structures of his embryonic polity.[3]

Other sources date Msiri's first successful deployment of flintlocks to a war against a group of Luba raiders from the Upemba Depression. According to Edgard Verdick—one of the founders of Lofoi, the earliest station of the Congo Free State in Katanga, in May 1891—the Yeke victory in the decisive engagement was primarily due to the "first flintlock muskets" (premiers fusils à pierre), the noise of which the Luba had never heard before.[4] This confrontation—which Verdick's chronology places immediately after the defeat of Lubabila, that is, c. 1870—is probably to be identified with, or set in the context of, Msiri's war against the Luba of Kilolo Ntambo, reference to which is made by Mukanda Bantu.[5] Telescoping what both Mukanda Bantu and Verdick imply was a longer period of time, the Plymouth Brethren missionary Frederick Arnot, in Garenganze between 1886 and 1888, wrote that when Msiri's party first settled among the Sanga, it had "four guns." Finding the then Mpande at war with the Luba, "Msidi came to the rescue of his father's friend, and after a few shots from his party, the Baluba, alarmed at the new weapons of war, took flight."[6] Arnot's story was later reprised almost verbatim by Joseph Moloney, a member of the European expedition responsible for the assassination of Msiri late in 1891.[7]

Contra the general tendency in most Yeke or Yeke-derived accounts to minimize the significance of firearms in southern Katanga in the 1850s and early 1860s, eastern Lunda historians state explicitly that no sooner had Msiri received permission to settle in Katanga from Mwata Kazembe VI Chinyanta Munona (1853/4–1862) than he began "to cause disturbances, and sent people off with guns to capture men as slaves."[8] British explorer Cameron, too, learned in 1875 that Msiri had originally entered Katanga "with a strong party . . . in search of ivory. When there he saw that his party, having the advantage of possessing guns, could easily conquer the native ruler. And this he forthwith proceeded to do, and established himself as an independent chief."[9] Cameron's data—gathered to the immediate west of the upper Lualaba River a mere twenty or so years after Msiri's arrival in the region—make it plausible to correlate Msiri's early career as mercenary and his subsequent usurpation of political power in southern Katanga with his having had access to significant firepower *ab initio*. Current oral sources support this view[10]—as does the fact that one of the words for "gun" used by the Yeke, *magoba*, appears to be a corruption of the Swahili *magobori*, the plural form of *gobori*.[11]

A further scrap of evidence in this direction is perhaps provided by one of the earliest war songs of the Yeke of Katanga. Collected in the late 1940s by Antoine Mwenda Munongo — Msiri's grandson and, from 1956, successor — "Wagya Magimu" ("May We Go to War") is said to have been composed by Msiri himself and to expatiate on one of the main motives behind his willingness to hire out his military services to local leaders: the need to secure wives for his unmarried east African followers and to beef up the number of his own clients and retainers. "Wagya Magimu" indicates that the Yeke war accoutrements always included a "horn powder flask" (corne à poudre). Once this was attached to its belt, it mattered little if the soldier was wounded or even killed, "because to obtain a young woman for one's house means to increase one's people like grasshoppers/locusts."[12] When pondering the likely role of guns in the early history of the Yeke conquest state, it is also worth bearing in mind that, by the middle of the nineteenth century, the approximate time of Msiri's establishment between the Dikulwe and Lufira Rivers, west-central Tanzania, his home area, had been acquainted with flintlock muskets for twenty or more years.[13] This being the case, it seems improbable that a Sumbwa caravan leader — one who, in his youth, might even have been employed as a porter by coastal traders[14] — would have ventured into the interior without availing himself of the most advanced possible military technology.

If the evidence relating to the place of guns in Msiri's initial conquests in the 1850s and 1860s is sparse, that pertaining to the trading connections through which he supplemented his arsenal in subsequent years is much richer. There is, indeed, no doubt that, having gained control of the districts inhabited by Sanga, Lemba, and related groups and resisted the eastern Lundas' attempt to dislodge him from his new power base, Msiri embarked on a systematic process of military modernization predicated on a more and more intense involvement in the long-distance trade with both the Tanzanian and, increasingly, the Angolan coasts. As will be seen in the next section, it was this same process that enabled Msiri to extend his sphere of influence over much of southern Katanga in the 1870s–1880s and to strengthen his polity's structures of economic domination.

From the beginning of his Katangese venture, Msiri was obviously in contact with Busumbwa and Unyamwezi, the Tanzanian regions whence he and most of his original followers hailed (see map 1.2) and where Katangese ivory and copper found a ready market.[15] In exchange for these, some muskets must have been obtained by the Yeke, for by the late 1860s these weapons—sometimes loaded with copper bullets made on the spot—formed a common accessory of the Nyamwezi copper traders whom Livingstone met near the Luapula River.[16]

The late 1860s may also have been the high point of Arab-Swahili commerce in southern Katanga. In 1868, the evergreen Said ibn Habib—whom we have already encountered in chapter 1—was reported to be trading in Katanga from his headquarters near the north end of Lake Mweru.[17] His colleague Muhammed ibn Saleh—whom Livingstone found among the eastern Lunda, where he had spent "twenty-five" years or more[18]—had also done business with Msiri. Indeed, it had been precisely such a trading relationship that had landed him in trouble with Mwata Kazembe Muonga Sunkutu. Suspecting Muhammed of complicity with Lubabila's murderers, the eastern Lunda king had ordered an attack against the trader and his followers.[19] The influence of Arab-Swahili long-distance traders in Garenganze appears to have declined in later years, possibly because of the eastern Lunda's hostility towards the Yeke, which made the eastbound trade routes from Bunkeya, the Yeke capital, dangerous,[20] or because of the coastal merchants' unwillingness easily to part with guns.[21]

Even though slave, ivory, and copper traders from both Zanzibar and the southern Swahili coast continued to visit southern Katanga until the death of Msiri (some of their representatives were reported in Bunkeya late in 1884, in 1886–1887, late in 1890, and as late as early 1891[22]), their role in the political economy of the Yeke kingdom was rapidly taken over by Luso-African and Ovimbundu caravans from present-day Angola. A direct trading link between Garenganze and the Ovimbundu plateau was inaugurated in about 1870 by João Baptista Ferreira, one of Silva Porto's associates, who responded to some earlier Yeke entreaties and supplied Msiri "with powder, guns, and cloth in exchange for ivory."[23] So briskly did the trade develop—and so significant did it eventually become—that it warranted explicit mention in Mukanda Bantu's aforementioned account.[24] Mukanda Bantu and other sources clarify that, unlike in the case of the eastern trade, the Yeke did not merely await the arrival of foreign merchants in their capital, but actively organized independent westbound caravans to transport the ivory and slaves accumulated by Msiri.[25] There are other clear proofs of the significance that Yeke elites attributed to the trade with Angola. By the mid-1880s, Msiri understood "very well" the "Umbundu dialect"—and so did many of Bunkeya's inhabitants.[26] Also, the Yeke Mwami had taken as his principal spouse the famous Maria de Fonseca, the Luso-African daughter of the Bihe-based trader Lourenço Souza Coïmbra. Maria was entrusted by Msiri with the running of the section of Bunkeya earmarked for visiting Angolan caravans.[27]

Already in 1875, Benguela, the terminus of Ovimbundu trade routes on the Angolan coast, was the principal port of entry for the guns (in primis, lazarinas) and gunpowder directed to Msiri. After negotiating Chokwe and Luvale countries and crossing the upper Lualaba to the south of the Upemba Depression,[28] "caravans commanded by half-caste Portuguese and slaves of Portuguese traders" traversed Sanga territory before finally attaining Msiri's capital on the Bunkeya River. Once in Bunkeya, heavily armed Angolan caravans were reportedly allowed to carry out slave raids in outlying districts. "On returning to [Msiri's] headquarters the slaves [were] divided between the traders and himself, in proportion to the number of guns furnished by his people." As a result of this, "large tracts of country [were] being depopulated."[29] By 1884, Paul Reichard remarked, the Ovimbundu-run trade in firearms and gunpowder had resulted in Msiri being equipped with an impressive force of "2,000 or 3,000 flintlock muskets" (2 bis 3000 Feuersteingewehren).[30] This arsenal exceeded that of any other regional power, and its "considerable" size was well known to neighboring peoples.[31] Without it, Msiri could not have pursued the aggressive foreign and economic policies described in the next section.

THE POLITICAL ECONOMY OF THE GUN

Having examined how Msiri and his immediate followers built up their arsenal over the years, it is time to explore the uses to which their guns were put. Here, my aim is to tease out the indispensability of firearms to the workings of the Yeke warlord state. Commercial imperatives dominated Msiri's policies and shaped his system of rule to a much larger degree than was the case in longer established polities, such as the Lozi's, the Ruund's or the Luba's, where the religious and ideological dimensions of royalty underlay its economic prerogatives. At the height of his power in the 1880s, Msiri acted very much like a merchant prince, lording it over Bunkeya—a bustling, cosmopolitan commercial hub located at the crossroads of several regional and long-distance exchange networks[32]—and exerting a pervasive control over the kingdom's external trade. No subject of the Mwami—the Portuguese explorer Ivens was led to believe—"dared to claim the right to own or trade" (ouse arrogar-se o direito de possuir ou negociar). "Unfortunate is he who tries, because the chief confiscates everything, without compensating the traders."[33] Such royal privileges were especially rigidly enforced in the case of the ivory and slave trades. Msiri's "extortions were manifold; thus every tusk of ivory had to be brought to the capital, and, if an official was suspected of concealing the precious material, woe betide him. In the dead of night the royal soldiers would surround the village; it would be reduced to ashes, and the people sold into slavery. The offender was lucky if he escaped with instant death, for Msiri delighted in diabolical refinements of cruelty."[34] There were, of course, solid economic reasons behind this severity. Not only were ivory and slaves exhaustible resources, they represented the most highly sought-after items of trade among Ovimbundu gun dealers. Thus, by centralizing in his hands the commerce in these two commodities, Msiri was also able to impose a practical royal monopoly over the acquisition and distribution of guns.

To be sure, the creation of dependencies and relations of reciprocity through the circulation of foreign goods was part of Msiri's political repertoire; yet such institutions of rule were largely confined to a small Yeke elite. Excluded from the economic and ideological devices with which Msiri sought to legitimize his supremacy in the eyes of his fellow Yeke,[35] the bulk of the kingdom's autochthonous inhabitants constituted a mass of primary producers to be preyed upon according to modalities that prefigured those that the Congo Free State (CFS) would later put in place. The methods through which regional resources were appropriated by Msiri reveal both the unprecedentedly exploitative and entrepreneurial nature of the Yeke system of rule and its reliance on the ready availability of imported muskets and gunpowder.

Msiri administered the heartland of the polity and even some of the more distant peripheral regions acquired over the course of the 1870s and 1880s by means of a network of territorial aristocrats, the *mwami mutemiwa*, and, especially, royally appointed officials, the *banangwa* (sing. *mwanangwa*). Distinguished by the right to wear the *libaka* copper bracelet and frequently chosen from within the ranks of Msiri's close patrilineal relatives, the Yeke and, more rarely, Yekeized banangwa served as tribute collectors and also as military and caravan leaders.[36] Insofar as ivory and slaves were concerned, the Yeke devised a "system of compulsory production quotas" destined to increase Msiri's access to such indispensable commodities.[37] Royal exactions caused real hardship to such subject peoples as the Lembwe, one of whose chiefs complained to Plymouth Brethren missionary Dan Crawford that Msiri had once given him "a paltry four yards of cloth," and "no remuneration whatever to his men," in return for a massive tribute of "forty teeth of ivory."[38] In another instance, a shortfall in the expected ivory tribute seems to have led to the execution of an unnamed local leader by the son of Dikuku, Msiri's brother and the Yeke mwanangwa among the Bena Ngoma of the *Mwashya*.[39] It is probably not coincidental that the future Sanga rebel leader, Mulowanyama, acted at one point as one of the "elephant hunters" at the service of the Yeke grandee "Molenga" (Mulenga, possibly to be identified with Kabobo, Msiri's cousin)[40]—or, perhaps, the time he was forced to devote to harvesting his high ivory quota made him appear so to the missionary Charles Swan.[41]

Given the resentment it generated, the quota system could only be enforced by the hated banangwa and traveling court envoys thanks to the hardware of violence that Msiri made available to them. It is indeed clear that Yeke settlements among subordinate groups were centers of significant military strength. "All" of the "male subjects" of Kimpoto ("Kimpotto") and Mirambo ("Milambo")—the first Yeke banangwa encountered by Delcommune to the north of Bunkeya in October 1891—owned "percussion-lock muskets" (fusils à piston), besides light swords with brass or copper decorations.[42] In a similar vein, in 1884, Capello and Ivens found firearms being widely used at Ntenke's ("N'tenque"), Msiri's representative among the copper-producing Lemba near present-day Kambove, who also insisted on receiving a Snider-Enfield breech-loading rifle and some cartridges from the two explorers.[43] Ntenke eventually turned against his erstwhile patron, exasperated as his subjects were by the "insatiable avidity of the king, who, besides ivory, also demanded slaves, men, women or children, whom he sold to the Biheans or the people of Kangombe."[44]

While benefiting from the ivory harvested by subject peoples and forwarded to Bunkeya by his territorial representatives, Msiri also encouraged the formation—and sponsored the activities—of hunting fraternities. As in the case of the banangwa, the arming of the members of these specialized guilds was the Mwami's responsibility.[45] Writing in the mid-1890s, CFS officer Clément Brasseur accused southern Katangese elephant hunters of risking their muskets (and their own lives) by using excessive quantities of gunpowder.[46] Ten years earlier,

Ivens had been less dismissive of local hunting skills, writing that the "war" (guerra) carried out against elephants by Msiri's semiprofessional hunters was so "fierce" (encarniçada) that the date of the pachyderm's "complete extinction" (completa extincção) in the region could not be far off.[47] Indeed, if Cameron is to be trusted, ivory was already becoming "scarce" in southern Katanga as early as 1875.[48] As on the upper Zambezi, then, a degree of temporary environmental degradation followed hot on the heels of the widespread adoption of firearms as tools for the production of animal capital.

Besides tribute, raids and wars were the principal means through which slaves were obtained for both trading and internal purposes. Prompted by the need to meet the demand for slaves—or the products of their labor—emanating from long-distance caravans, slave forays punctuated the life of Garenganze and accounted for the spread of violence and overall militarization of society in the region in the latter part of the century. "Large numbers of slaves [were] brought into the capital every year by returning war parties, and [were] sold chiefly to Arab traders from Zanzibar and to Ovimbundu traders from Bihé. Strong young men have been sold for ten or twelve yards of cotton cloth."[49] Women were perhaps more frequently kept in Bunkeya and surrounding areas with a view to exploiting their labor and reproductive potential. According to Arnot, "an immense number" of captive women resided in the heartland of Garenganze, "so that the proportion of women to men [was] very unequal; consequently polygamy [was] carried on to a shameful extent."[50] The pervasiveness of slave raiding was no doubt one of the reasons behind Bunkeya's oft-remarked cosmopolitan nature. For "[w]hat means the babel of tongues heard in this country, if not that Msidi's war parties have brought in from nearly every point of the compass gangs of poor downtrodden mortals, who were swooped down upon in their little hamlets far away and carried off?"[51]

Though foreign traders are said to have occasionally taken part in Yeke forays (see Cameron's remarks, above), most slave raids were carried out by relatively small but well-armed Yeke war bands. Since these military incursions targeted primarily the periphery of the state, it seems that Msiri sometimes found it difficult to limit their range and "[restrain] his soldiers from extending [them] mercilessly, when once he has banded them together to attack any chief."[52] According to Legros's informants, the average Yeke raiding party was led by one or more *mutwale* (who were also frequently *banangwa*) designated and appointed by Msiri. Each mutwale commanded a *bulungu*, the basic Yeke military unit. This consisted of a core of about twenty gunmen, whose firearms were personally entrusted to them by Msiri, and a further group of lightly armed fighters, numbering between sixty and one hundred.[53] Depending on the nature of the particular expedition being prepared, several bulungu could operate together. In 1885, for instance, Capello and Ivens encountered a large Yeke armed band near present-day Lubumbashi. On this occasion, the overall leader of the "horde of brigands" (horda de salteadores), Dikuku ("Licuco"), Msiri's brother, commanded "about 400 men" equipped with "firearms, arrows, [and] spears" (armas, fleches, zagaias).[54] By the late 1880s, Msiri's fighters were no longer equipped solely with lazarinas. Unlike in southern Africa, breech-loaders remained a rarity: as late as 1890, not even Maria de Fonseca knew how to operate one.[55] But percussion-lock muskets had become more common than they appear to have been at the time of Reichard's passage. Shortly after his arrival in Bunkeya in 1888, for instance, Swan discovered that the only buyers for his stock of caps were "Msidi and his warriors." As soon as he understood that they "want[ed] them principally for their plundering expeditions," he resolved to stop dealing in them.[56]

The available written evidence does not clarify how the Yeke military was recruited. Small semiprofessional militias were attached to each mwanangwa;[57] it is, however, also possible that an element of compulsion was present. In 1887, Arnot stumbled across a beheaded corpse in Kimpata, the quarter of Bunkeya supervised by Msiri's first wife, Kapapa. He then learned that the execution had been ordered by the Mwami himself, who "had caught this young man skulking." "It seems that Msidi had ordered him some days before to join a war party that had already gone out, and as it was his third or fourth offence Msidi ordered his immediate execution."[58] Yeke raiders, on the other hand, could look forward to being rewarded for their services: "the fighting has become fierce"—went a Yeke song. "It matters not. We shall go to war anyway, because we shall be clothed by our king, by our suzerain."[59]

Full-blown wars of expansion, such as those that took place against peripheral Luba leaders in the Upemba Depression from the 1870s or along the Luapula River from about 1880, were certainly less frequent than smaller-scale predatory raids. They, however, were no less informed by economic imperatives and dependent on adequate military supplies. Arnot, for one, thought there was a clear correlation between the consolidation of the Ovimbundu trading connection from c. 1870 and Msiri's subsequent policy of "aggressions upon the surrounding tribes."[60] If guns influenced the pace and timing of Yeke expansion in the 1870s–1880s, the uneven spread of the new technology in Katanga had something to do with the mixed fortunes of these military undertakings. In general, the Yeke scored resounding successes against lightly armed opponents in or near the heartland of Garenganze. Their superiority, however, was less clear against people with a long familiarity with firearms—most notably, the eastern Lunda, whose first exposure to muskets dated to the first half of the century (see chapter 1). In the early 1880s, Yeke armies overran the western half of the eastern Lunda heartland, inaugurating a violently extractive system of occupation that prompted mass flights across the Luapula River and that has left vivid memories in the area.[61] Yet, this initially successful conquering thrust along the lower Luapula was followed by a military stalemate, characterized by a series of desultory cross-river raids by the two warring parties and by the Yeke involvement in the ongoing Kazembe civil wars.[62] Central to Lunda resistance was the attested ability of Kanyembo Ntemena, the eastern Lunda king from 1885/1886, to increase the firepower at his disposal. In 1883, his predecessor, Mwata Kazembe IX Lukwesa Mpanga, had been reported to have a mere 120 guns (including the 49 percussion-lock muskets and the one Gras breech-loader that he had recently confiscated from the French explorer Victor Giraud).[63] Seven years later, however, Kanyembo was already described by Alfred Sharpe, the envoy of the British South Africa Company, as "the most powerful native Chief [he had] yet seen in Africa."[64] In the late 1890s, a Company official who had earlier clashed with the still unconquered eastern Lunda "warned his superiors that 'Kazembe can turn out a large number of men (over 1,000) armed with muzzle loading guns'" imported from the east coast.[65]

To foreground the military significance of guns in southern Katanga is, of course, not the same as denying the importance of other factors, such as terrain and tactics. This is aptly borne out by the performance of the Luba of Kinkondja during what Thomas Reefe termed the "Wars of the Upemba Depression." Ngandu, the ruler of Kinkondja in the early 1880s, was one of the few Luba peripheral leaders effectively to thwart the Yeke despite lacking firearms.[66] In the wars against Kinkondja, in which Msiri was personally involved for several years,[67] inadequate knowledge of local environment and the tactics of amphibious warfare sapped whatever advantage firearms might initially have given Yeke dry-land soldiers.[68] Other gun-poor communities, such as some Lomotwa and Bena Mitumba groups, took advantage of the mountainous nature of their homelands, using the protection afforded to them by a network of secret caves and dens to retain a degree of independence vis-à-vis the Yeke and avoid some at least of the spoliations that the latter committed elsewhere.[69]

As befits the members of a gun society, Yeke men of violence were well aware of the different properties and performances of various models of guns. In one popular song, for instance, Msiri's soldiers are depicted as clamoring for newer guns, since "flintlock muskets are decimating us" (les fusils à silex nous font décimer).[70] Local blacksmiths produced edged weapons, if so required by Msiri. "The king gave the order, let's forge some weapons for him" was the refrain of one of the labor chants collected by Mwenda Munongo.[71] They—not unlike upper Zambezian artisans (see chapter 2)—could also manufacture iron and copper bullets and, apparently, "repair all the parts of a gun."[72] The proficiency of some Ushi ironsmiths working for Msiri was brought to the attention of Captain Brasseur in the mid-1890s and is still remembered in present-day Bunkeya.[73] This is not entirely surprising, since the leading Ushi potentate of the day, Milambo Myelemyele, was a renowned gun-mender, whose "real skill" (habileté réelle) had much impressed Giraud in the early 1880s. The same Giraud also left us the following admiring description of the obsolete, but clearly effective, equipment with which Milambo plied his trade: "old flintlock plates, old files, filletless screws of various sizes—all dating to at least the day of the invention of firearms. One day, I saw him making a complete wooden stock, which—my word!—was very well suited to the barrel to which it was destined."[74]

Yet it may be surmised that the size and frequency of gun imports at the height of Ovimbundu/Yeke trade were such that this set of technical competences was not honed to its full extent. The spur to develop systems for the local production of gunpowder might have been similarly reduced by abundant foreign supply. The historical importance of European-made trade gunpowder and the other imports with which it was associated in southern Katanga is borne out by a "cursing ceremony" performed by Ovimbundu caravaneers in the mid-1890s, the time in which they were being squeezed out of Garenganze by the Congo Free

State (see chapter 4). Upon reaching the upper Lualaba, and before deviating from the eastward itinerary that had earlier taken them across the Lufira and onwards to Bunkeya, they paused and, looking east, poured out "invectives into the air." "A powder barrel is ominously filled with sand to remind them of the kind of powder (!) [sic] the people here had before the Bihéans had mercy on them and brought them the European kind. This barrel of sand is put across the road to the east . . . and then an old bow and blunt arrow are laid on the top of it, eloquent with irony as to their primitive weapon, and then, topping all, comes a worn-out antelope skin—the climax of insult as to their old-time nakedness."[75]

Among the Gwembe Tonga of the middle Zambezi River, the art of making gunpowder was imported by Chikunda "transfrontiersmen" (see chapter 1). Both the Chikunda and the Gwembe Tonga manufactured serviceable gunpowder "by taking the bark off the *twetwe* tree and combining it with either saltpeter or soils rich in acid from rabbit urine. They then burned the mixture and extracted the powder from the charcoal remnants."[76] Further north, in the vicinity of present-day Lusaka, "calcareous tufa" was used that contained "potassium nitrate" (saltpeter) from the excrement of baboons, and an extensive manufacturing complex was still operational in the 1910s.[77] This African-made gunpowder was traded relatively widely, with some of it finding its way to the gun-rich Kaonde, who had a special name for it: *mapikwa*.[78] The Yeke never did evolve such techniques—similar in this to the societies of the Angola-Zambia border, among whom there is little or "no evidence of any indigenous gunpowder-making industry."[79] Unlike the peoples of the upper Zambezi, however, the Yeke, a small immigrant elite who had established themselves by the gun and whom guns enabled to face up to widespread local hostility, were to pay a high price for their dependence on external military imports.

Given the density of the narrative presented above, it is perhaps in order to summarize our conclusions so far. Foreign Yeke conquerors— this and the previous section have argued—relied on extreme force and firepower to establish, maintain, and expand their structures of politico-military and economic domination over southern Katanga. By means of gun-induced violence or intimidation, Msiri, his inner circle, and their militias were able to obtain the commodities (principally, though not exclusively, ivory and slaves) that fed the long-distance trade with the east and, especially, the west coast of the continent. Up to a point, the cycle was self-perpetuating, for such commodities were employed to obtain more guns for the further production of tradable goods. However, the fact that the workings of the Yeke pillaging economy were predicated on the continuing availability of an imported technology implied a serious structural weakness. Ivens noticed as much in 1884, pointing to the "very difficult situation" (mais complicada situação) in which Msiri would find himself if the Benguela markets were to be suddenly closed to him by circumstances beyond his control.[80] As will be seen, there was a touch of prophecy in the Portuguese explorer's remarks.

REBELLION AND COLLAPSE

Traffic remained heavy on the Bunkeya-Bihe route in 1889–1890,[81] when Msiri's still "large supplies"[82] of imported guns and gunpowder enabled him temporarily to prevail over the coastal raider Shimba and his heavily armed following, who had earlier settled on the southwestern shores of Lake Mweru and whom an expedition led by Mukanda Bantu eventually forced to withdraw to an impregnable stronghold on Kilwa Island in the spring of 1890.[83] The economic and military situation of Garenganze, however, took a precipitous turn for the worse from February 1891, the month that witnessed the outbreak of the Sanga rebellion—a product of the oppressiveness of Msiri's rule and his incapacity to legitimize his position in the eyes of the Sanga and other autochthonous peoples. The insurgency's spark, as is confirmed by several coeval sources, was the accidental murder of Masengo, a Sanga woman, and the subsequent refusal on Msiri's part to hand over the legal culprit—that is, the Yeke owner of the slave who had fired the accidental gunshot.[84] But the fate of Masengo was a mere *casus belli*, behind which, as the eyewitness Crawford put it, lay "deep, deep-seated grudges of long-standing against Msidi's tyranny" and the exploitative economic system that revolved around his person and his gunmen.[85] Mukanda Bantu himself confessed to Alexandre Delcommune that he understood the Sanga rebels, who had suffered from Msiri's "terror" (terreur) and who, "despite everything, regarded him as a mere stranger" (malgré tout, ne voyaient en lui qu'un étranger).[86]

Led by Kalunkumia, Mpande Kya Mulemba's successor, and his "sons," the Sanga, who had certainly managed to accumulate some muskets by the time of the rebellion, began a series of nightly attacks against Bunkeya and other centers of Yeke power.[87] The first recorded action of the Sanga "guerrilla war" (the expression is Crawford's[88]) took place on 21 February 1891, when seven huts in Msiri's own section of Bunkeya,

Nkulu, "were maliciously set on fire and burnt to the ground." Two women and a child were shot in a neighboring quarter of the capital, and one more casualty was recorded in Kankofu, the neighboring village of mwanangwa Ntalasha ("Ndalasia"), Msiri's classificatory brother.[89] Thereafter, Sanga forays increased in intensity and geographical scope: during the night between 25 and 26 February, "two or three houses were burned down . . . at different villages."[90] Between March and April, the Yeke colonies of both Dikuku ("Lukuku") and the late Mulenga came under attack.[91] By then, Msiri "no longer [slept] unprotected, but ha[d] a large body-guard sleeping around his house every night,"[92] and the inhabitants of Bunkeya's outlying quarters were being forced to seek shelter in the central, fortified areas of the capital.[93] Even though one of the obvious aims of the Sanga guerrillas was to generate panic among the inhabitants of Yeke settlements, their attacks were not always random. On occasion, they targeted specific members of the Yeke or Yekeized hierarchy, such as Muluwe, the *mpande*-wearing "headman" who was "shot through the heart" in the early hours of 10 May.[94]

Under normal circumstances, Msiri's war machine would have enabled him quickly to put down the insurgency. The Sanga knew it and thus resolved to pair their direct military actions with a policy of economic strangulation destined to sever the lines of communication between the heartland of Garenganze and the Ovimbundu plateau. By imposing a blockade on the Lualaba ferry and the route between it and Bunkeya, the Sanga of Mulowanyama—whose stronghold lay to the west of Bunkeya, near the Dikulwe River, and who would soon be joined by the renegade Ntenke's Lemba—prevented Msiri from resupplying his military stock with new muskets and, especially, gunpowder, at a time in which his forces were also being severely tested by the eastern Lunda, on the lower Luapula, and by Shimba, on Lake Mweru. The Sanga tactic proved highly successful in restricting freedom of movement in southern Katanga, as attested early in June by Paul Le Marinel, who found it impossible to travel between Bunkeya and Kambove because of "the Sanga blocking the routes" (les Basanga bloquant les routes).[95] In so doing, the Sanga turned the nightmare scenario first imagined by Ivens in 1884 into a reality.

Finding himself constrained by lack of gunpowder, Msiri could not reply decisively to the outbreak of the rebellion and was forced to postpone his counteroffensive for several months.[96] The Yeke supply situation registered a temporary improvement in May, thanks to the powder given to Msiri by Le Marinel ("5 *barils*") and by a well-stocked Angolan caravan that had managed to force the blockade.[97] Finally "in a position to make a big muster of war parties," Msiri dispatched two separate military expeditions in the early summer of 1891. While Dikuku went up the Lufira with a view to protecting the *Mwashyas'* salt pans from attacks by the Sanga of Mutwila, another column, led by the warrior-prince Mukanda Bantu, took a southwesterly course and confronted the Sanga at Kalunkumia's. After much firing, Kalunkumia's "Va-sanga were compelled to evacuate the town and flee south, in the direction of Ntenke's. No chase seems to have followed, and the war parties returned to the capital without striking any decisive blow."[98]

Despite this setback, the Sanga and their growing network of allies held out, while the internal situation of the besieged and isolated Garenganze deteriorated rapidly. Not only had the "war . . . reduced Msidi's ivory tribute this year to almost nothing," but continuing Sanga forays also forced the inhabitants of Bunkeya not to "venture out far to the fields they were wont to cultivate" before the start of the rebellion and to limit themselves to tilling "the poor stony soil" of the hill around which the Yeke capital had been built.[99] This had the effect of accelerating the onset of a famine brought about by that "year's exceptionally dry conditions" (sécheresse exceptionnelle de l'année).[100] Famine and ongoing security concerns, in turn, led to population losses. By October–November, "more than half" (plus de la moitié) of Bunkeya's population had fled,[101] and Msiri was again short of powder.[102] Dikuku, meanwhile, had failed to overcome Mutwila,[103] and the Lomotwa under the yoke of mwanangwa Kifuntwe ("Kifumtiè") had thrown their lot in with the rebels.[104] Running out of options, Msiri sought unsuccessfully to induce both Delcommune and the CFS officials at Lofoi to fight the insurgents on his behalf, and he might have even contemplated the possibility of shifting the location of his capital.[105] By mid-December, when William Stairs and his three hundred men (two hundred of whom were equipped with modern breech-loaders) made their entry into Bunkeya, the famine was "appalling" (affreuse) and the central symbol of Yeke power a pale shadow of its former self. The Angolan trader Coïmbra told Stairs that the ten sections into which Bunkeya had been subdivided a mere three years earlier were now reduced to one and that the hills to the southwest of the town, once "dotted with flourishing villages" (couverts de villages florissants), were now almost entirely depopulated.[106]

Lack of gunpowder and the Lualaba blockade remained Msiri's most pressing concerns to the very end. One last Yeke sortie, against Mulowanyama, took place between November and December, in retaliation for an earlier attack on a party of eastbound traders at the Lualaba ferry. At the time, the missionary H. B. Thompson was being detained by Mulowanyama. The latter—who appreciated the neutral stance taken by the Plymouth Brethren and their refusal to assist Msiri in communicating with "traders to bring powder to him"—was ready to give Thompson permission to proceed to Bunkeya, but insisted that the members of his caravan who carried powder be forced to stay put until they had exchanged all of it locally.[107] On 2 December, while negotiations were still ongoing, Thompson witnessed Mukanda Bantu's assault. "I saw men and women returning from the fields in hot haste, and making for their villages beside our camp. Immediately I heard the beat of a war drum, and a number of shots were discharged in the bush, about 250 yards from our camp. . . . Msidi's party caught about eleven of the Va-sanga in the fields and forthwith decapitated them, as a *quid pro quo*, I suppose, for those killed at the ford by the Va-sanga." Thompson's intervention apparently prevented further bloodshed and convinced the Yeke aggressors to decamp.[108] Even the Yeke military no longer believed in victory. Far from being the springboard for an improbable *reconquista* of southern Katanga, the desultory skirmish described by Thompson proved to be last roar of the dying lion.

The explorer-*cum*-mercenary Stairs clashed with Msiri over his determination to force the CFS flag on the Yeke Mwami and refusal to furnish the latter with the gunpowder he craved.[109] Tension escalated between 16 and 20 December, the day on which Msiri, in a burst of offended dignity, threatened Omer Bodson with a sword he had earlier been given as a present, and was as a result shot dead by the Belgian officer and his escort.[110] It is of course fitting that Msiri—whose political career had been so thoroughly shaped by the advent of firearms that he had reportedly seen fit to call one of his children "Chifamina-Chamundu," or "the two bullets in the gun" (les deux balles dans le fusil)[111]—should die riddled by European bullets.

↪

The negligible resistance put up by the Yeke of Bunkeya following the demise of their leader underscores the extent of the process of disintegration that had taken place over the previous few months. At a more profound level, it illustrates the fragility of the bases on which Msiri's warlord state had been built. Because of the circumstances attending upon the inception of Garenganze, technology in the form of imported firearms and gunpowder was deeply interwoven with the life of the polity, favoring its inception and subsequently underpinning its administrative structures and economic policies. Flintlock and percussion-lock muskets, in this context, were *the* Yeke means of military domination and commodity production. Yet, for all the intensity of the Yeke engagement with them, firearms could not secure the stability of Msiri's creation. Lacking legitimacy vis-à-vis subject groups, and drawing heavily on superior firepower to keep its predatory policies in place, Garenganze proved vulnerable even to a temporary interruption in the flow of military supplies from the west coast. In demonstrating that technological superiority alone constitutes an insufficient basis for the long-term survival of any system of domination, the trajectory of the Yeke warlord state provides further proof of the inadequacy of deterministic understandings of the relation between technology and society. The Sanga rebellion, in some ways, can be conceptualized as a reassertion of the rights of society over those of technology.

Following the death of Msiri, the Yeke of Mukanda Bantu—the son of Msiri whom Stairs appointed as the new Mwami on 29 December 1891[112]—were surrounded by murderous popular hostility and faced with the prospect of complete annihilation. Mulowanyama— who, like most other Sanga, was animated by a "most vivid" "hatred" of Yeke "looters"[113]—is said to have appealed to Stairs to "exterminate" all of Msiri's people.[114] In some cases, these murderous intentions were put into practice. In June 1892, for instance, the Lomotwa of the then Mufunga assassinated their former mwanangwa, Kifuntwe—an episode which Crawford read as a demonstration of the fact that "they who once were the oppressors are now in a degree the oppressed."[115] In this context, Mukanda Bantu and his few remaining followers were left with no choice but to rally behind the newly arrived whites, becoming the latter's most dependable local allies in the course of the brutal "pacification" of Katanga promoted by the CFS from 1894. Fighting as auxiliaries to the Force Publique against the Sanga and other peoples who had initially greeted the Europeans as liberators, the Yeke ensured their continuing access to firearms. By thus shaping from below the violent European conquest of Katanga and the early workings of the CFS, they were able to partly rebuild their threatened regional hegemony.

4 ⊸ Gun Societies Undone?
The Effects of British and Belgian Rule

BY TRANSFORMING the overall sociopolitical context, colonialism altered the terms of African engagement with firearms, but the process was both long-drawn-out and complex. The central savanna was partitioned between separate colonial powers and systems of rule, each underpinned by specific economic policies and local patterns of alliance and/or confrontation. Some central African societies negotiated their inclusion into emerging colonial structures. In other instances, Euro-African cooperation was, literally, forged in blood, with African men of violence making a decisive contribution to the colonial takeover. Finally, other communities either resisted the Europeans or, perhaps more commonly, sought to evade their attention for as long as possible. Because of this unevenness, the relations between Africans and firearms did not change everywhere at the same pace or according to the same criteria. In this regard, the contrast was pronounced between North-Western Rhodesia and the Congo Free State, the two colonial territories into which the gun societies examined in chapters 2 and 3 were incorporated.

In October 1889, Cecil Rhodes's newly formed British South Africa Company (BSAC) was granted a Royal Charter of Incorporation. In return for extensive economic privileges, the BSAC undertook to administer on behalf of the British government a wide portion of south-central-African territory, the precise limits of which were to be determined by means of the stipulation of "treaties of protection" with local rulers. One of Rhodes's chief aims to the north of the Zambezi was to secure Katanga and its rumored mineral wealth.[1] Once such plans foundered in the early 1890s, the BSAC found itself saddled with a vast territory, the economic potentialities of which were taken to be limited. The slow pace and (partly) negotiated nature of the occupation of what became known as North-Western Rhodesia must be understood in this context.[2]

The Congo Free State—within the boundaries of which Katanga was placed—was a different kettle of fish. Born out of the ambition and diplomatic maneuvers of the Belgian king, Leopold II, the CFS was administered as his personal property until 1908, the year in which it was transferred to Belgium as a full-blown colony. Initially formally split between the state itself and the Compagnie du Katanga, the administration of Katanga was taken over from 1900 by the Comité Spécial du Katanga, a concessionary company in which the CFS held majority share. Such legal intricacies made hardly any difference to most Africans in southern Katanga, where the state's determination to enforce its ownership of the Congo's wild products and control over the labor of its inhabitants became the dominant fact of life from as early as the mid-1890s.[3]

The delayed imposition of colonial rule in North-Western Rhodesia meant that precolonial patterns of trade and gun domestication lasted well into the twentieth century, and that, for several years, Lozi, Luvale, and Kaonde continued to understand and deploy firearms as they had done before the Scramble for Africa. Guns and game laws were a long time in the making in North-Western Rhodesia, but when they did finally come about in the early 1920s, their impact proved considerable, leading to the weakening and atrophy of long-established local gun societies. Conversely, colonial rule in southern Katanga was imposed both comparatively quickly and brutally. However, because of the nature of the relationship between the CFS and the Yeke and the latter's implication in the violence of the former, the disarmament of the late Msiri's people was never contemplated, and the colonial conquest did not bring about a radical reconfiguration of their understanding of firearms as tools of military and economic hegemony. The incorporation of Yeke territory into the emerging Katangese industrial economy in the early twentieth century was to have longer-lasting consequences. But even this process did not lead to the complete undoing of Yeke gun society, within which firearms were now increasingly recast as individually owned hunting tools, to the detriment of their

hitherto predominant application as centrally controlled means of political domination and commodity extraction. More consistent with the
zeitgeist, such uses mirrored earlier patterns of technological domestication near the sources of the Zambezi and resulted in the gun retaining a foundational role in what was left of Yeke society and polity all the
way to independence and beyond.

GUNRUNNING IN NORTH-WESTERN RHODESIA

Unlike in southern Africa's settler colonies, colonial rule and African
disarmament did not immediately go hand-in-hand in North-Western
Rhodesia, where the unlicensed importation of guns, powder, and ammunition was forbidden in 1901,[4] but where ownership of firearms by
Africans remained initially unregulated. This was a result of the BSAC
having asserted its legal rights over the country by engaging in treaties
with the Lozi king, Lewanika (see chapter 2), between 1890 and 1900.[5]
The circumscribed judicial independence granted to the Lozi under
the terms of these successive agreements meant that the Company did
not deem it fit to seek to push through a comprehensive gun legislation
in the 1900s.[6] Early local experiments in the voluntary registration of
guns did take place in select North-Western Rhodesian localities, such
as the Batoka district from 1903.[7] But these uncoordinated initiatives
did not amount to a coherent effort at gun licensing or, even less, to an
attempt at enforcing such near-universal African disarmament as that
brought about by the unyielding North-Eastern Rhodesia's "Fire-Arms
Restricting Regulations (Natives and Asiatics)."[8] Game laws were similarly skewed in favor of North-Western Rhodesia's Africans. Whereas
their North-Eastern Rhodesian peers had been expected to take out
hunting licenses from as early as 1900,[9] Africans within the boundaries
of North-Western Rhodesia were explicitly exempted from the stipula-
tions of the first game preservation regulations to be issued in the country in 1905. This privilege, too, was "in accordance with the provisions
of the Concession granted by King Lewanika Paramount Chief of the
Barotse Nation to the British South Africa Company dated October
17th 1900."[10]

The absence of internal regulations concerning African gun ownership and hunting turned gunrunning from Portuguese Angola into
one of the defining features of the early colonial period in North-
Western Rhodesia. A number of additional factors help explain the size
of the phenomenon. While the demand for captive labor in Angola

remained high long after the formal abolition of slavery,[11] Portuguese officials struggled to establish even the barest form of administrative control over the sprawling territory that made up the central and eastern regions of the colony.[12] They, moreover, were strongly suspected of having a stake in an illicit cross-border trade that was also fueled by the initially unrestricted importation and sale of flintlocks, percussion-lock muskets, and gunpowder in the country under their charge.[13] Finally, Ovimbundu and Luso-African smuggling activities were greatly facilitated by both the extent of the frontier between Angola and North-Western Rhodesia and, especially, the uncertainty surrounding its definition until 1905, the year in which the king of Italy was requested to adjudicate the long-standing boundary dispute between the British and Portuguese governments. This meant that throughout the early 1900s Mambari traders could continue to operate in North-Western Rhodesia, safe in the knowledge that all that was required to avoid prosecution and/or the confiscation of their merchandise was to beat a quick retreat to such contested border areas as BSAC patrols dared not encroach upon for fear of causing a diplomatic incident.[14]

As in the nineteenth century, Barotseland's wealth remained a powerful magnet for Angolan traders, some of whom were reported to be buying cattle in Lealui, the royal summer capital, in June 1900 in exchange for "gunpowder, arms and calico."[15] Another "large caravan . . . from Bihe" made its entry into Bulozi a few days later, prompting the acting British resident to voice his "apprehension" at "the constant increase of the importation of arms and ammunition from the West Coast to this country."[16] At first, the impact of the import regulations of 1901 was clearly negligible. In 1903, Lewanika—who still used the allocation of firearms and ammunition as a means to bolster his position vis-à-vis an officialdom which was bearing the brunt of the BSAC's steady encroachment upon many of its former prerogatives[17]—was able to persuade the administration to order as many as four thousand Martini and other cartridges for distribution among his trusted councilors. The Company acceded to the king's request, for it expected that this "very marked concession" would "stop all illegal traffic in powder and cartridges."[18] In fact, seemingly condoned by Lewanika,[19] gun smuggling remained common,[20] though the boundary award of 1905 finally equipped BSAC officials to deal with it more effectively. The "half-caste" trader Ferreira was one of the first traffickers to experience the fast-improving boundary-policing capacity of the BSAC in

FIGURE 4.1 Ovimbundu trading caravan, c. 1890. Source: Frederick S. Arnot, *Bihé and Garenganze; or, Four Years' Further Work and Travel in Central Africa* (London: James E. Hawkins, 1893), frontispiece. Note the number of guns in the possession of traders and porters. The first man on the right is holding a lazarina, recognizable by its characteristic grooved stock.

Barotseland. First expelled from the district when found in possession of "65 guns and a large supply of ammunition" in July 1911,[21] he had his stock (which included twenty-five flintlocks) confiscated and his large camp burned down when he was apprehended for a second time inside British territory a few months later.[22]

The gradual—and never entirely complete—closing down of Barotseland from 1905 did not pose insurmountable problems for Mambari gunrunners, who were also very active in the still almost entirely unpoliced territory to the north, where, of course, they boasted long-established trading links with the Luvale, whose raiding economy and masculine identity their lazarinas had helped to beget over the course of the previous century. Among the Luvale of the *Kakenge* and the *Nyakatoro*, the slave trade was still thriving c. 1900, with "a gun or 40 or 80 yards of calico" being "the purchase value of an adult slave."[23] Rubber exports were also being paid in guns and gunpowder, which Luvale men bought "far more readily than ordinary trade goods."[24] Undoubtedly many of these guns also found their way to the Luvale and Lunda inhabiting what would become the Balovale subdistrict of Barotseland in 1907–1908. Writing several years after the events, Native Commissioner (NC) J. H. Venning, the first official in charge of Balovale, still remembered vividly the "large numbers of guns" owned by "both the Malunda and Malovale," and the troubles he faced in bringing illegal imports from Angola under control.[25] Venning also claimed to have witnessed "the last slave raid" to occur in the area.[26] This may have been so, but there is no doubt that Mambari gunrunners remained a force to be reckoned with for some more years to come. A large trading outpost in a fortified Luvale village, for instance, was discovered in 1911. Venning's successor surmised it must have been "established there quite a long time."[27]

The trade in slaves and guns was also much in evidence in neighboring Mwinilunga, where, again, no administrative work took place until the beginning of 1908, when the Balunda subdistrict of Kasempa district was got off the ground by NC Bellis. Mambari dealers had certainly been at work in the area at the end of the nineteenth century,[28] and they continued their frequent visits throughout the 1900s. In 1904, the then Kanongesha, the Lunda-Ndembu paramount, "had seven or eight slaves for sale and was expecting the arrival of Mambari purchasers."[29] Two years later, Copeman, the Kasempa district commissioner (DC), and an escort of Barotse Native Police traveled to within fifty miles of Mwinilunga with a view to intercepting "certain Portuguese traders" who had been reported to be "trading guns, powder and caps in return for slaves, ivory and rubber." Three "stores" were eventually located. A "large quantity" of forbidden goods was discovered, while the "many empty powder canisters which were found in the different stores and lying about the camps testified to the large trade which has been going on in this commodity."[30] One of the ways in which G. A. McGregor, Balunda's first full DC from June 1908, sought to justify his deliberate "policy of violence"—a policy for which he was eventually removed from his post after only one year in charge—was to bring to his superiors' attention the "disturbed condition of affairs [which] existed" in the district, partly as a result of the actions of "the slave dealing parties of Mambunda and others who for many years past have raided our Territory from Portuguese Angola."[31] Being the target of an official inquiry into his conduct, McGregor had an obvious interest in painting an inflated picture of Ovimbundu and Luso-African smuggling, but even his successor Bellis, who had himself been shot during his earlier stint as NC of the subdistrict early in 1908, thought that "the number of guns and quantity of powder in the possession of the natives" of Balunda posed a "considerable personal risk" to his staff.[32] As late as 1910, having "suffered much from slave raiding in the past," the Lunda to the west of the Kabompo River were still said to regard "the possession of guns and gunpowder [as] a necessary protection . . . To obtain gunpowder they will do anything, and in obtaining it they often get into the hand of half-caste or alien native traders in Portuguese territory where large supplies of gunpowder are bartered to them."[33]

As has been argued in chapter 2, the Kaonde of the *Kasempa* had responded enthusiastically to the long-distance trade in the latter part of the nineteenth century, incorporating Ovimbundu-imported muskets into their hunting economy and endowing them with the additional function of units of exchange. By 1901, their area was "flooded with guns and ammunitions"[34]—a state of affairs that prompted the BSAC to inaugurate a police post at Fort Kasempa in 1902.[35] In the face of Mambari obduracy, however, the task of controlling the district proved far from straightforward. In 1903, Sub-Inspector Macaulay intercepted and dispersed two Mambari caravans between Kasempa and the Congo Free State border.[36] It was probably on this occasion that he confiscated "about 50 guns, 100 bags of gunpowder, 50 loads of slave trading goods and three women, one of whom actually had the slave shackles on her neck."[37] Another Mambari camp was surprised near the then Mushima's village at about the same time. In this instance, only three guns were destroyed and fourteen bags of gunpowder impounded.[38] One of the problems faced by local administrators in dealing with Angolan smugglers was that the "natives will give no information [as to] their whereabouts." This was scarcely surprising, given that the Mambari were the Kaonde's principal suppliers of firearms and that, by now, "every native ha[d] a gun and a great many . . . two or three in their possession." It was these same guns, Macaulay surmised, that accounted for the considerable quantity of ivory still exported from the district and the presumed decrease in elephant populations.[39] Even the immediate neighborhood of Fort Kasempa was not free from Mambari activity. Being "kept so well informed by the natives . . . it must be quite by chance if one comes in contact with them."[40]

In the early 1900s, Bwila and Lenjeland were still common destinations for the Angolan slave dealers operating in Barotseland and the Kasempa district.[41] To be sure, guns remained rare among the Ila of the middle Kafue both before and after the formal imposition of European control. As late as 1899, James Stevenson-Hamilton was much impressed by the quantity of game near present-day Namwala, which he attributed to the "natives [having] no guns."[42] "The Abashakulumbwi [Ila] inhabiting the Kavuvu [Kafue]"—noted Val Gielgud, the first

BSAC representative in the area, in 1900—were "armed with assegais and spears only. I saw no guns or bows among them."[43] Lenje country, however, had changed beyond recognition since the early 1850s. The Lenje had clearly fallen in line with developments elsewhere in North-Western Zambia and were now widely employing guns for both hunting and raiding purposes. Once more, the gun frontier was being driven forward by Mambari traders, whose growing centrality to the Lenje economy topped the list of Gielgud's worries. "Owing to the enterprise of the Mambundu traders," he reported following a visit to the Lukanga swamp late in 1900, "nearly half" of the villagers he had seen possessed guns.[44] In the course of his patrol through Lenje country, Gielgud and his Ndebele police intercepted and fought a Mambari party at Kasonkomona's ("Siasonkomola's") on 2 October. Although a good many of the loads dropped by the fleeing traders were looted by locals, Gielgud still "succeeded in obtaining and destroying about 30 muzzle loading guns and 100 lbs of gunpowder." A second Ovimbundu slave caravan was surprised two days later. Again, despite the "Abalengi" helping themselves to "large quantities" of abandoned trading goods, Gielgud managed to seize and destroy "about 18 guns and 150 lbs of powder." By this time, apart from procuring arms and gunpowder at a fast rate, the Lenje had also begun to manufacture iron bullets.[45] All in all, his experiences with the Lenje convinced Gielgud that a process of wholesale rearmament was afoot and that "one of the most beneficial measures that could be taken in this country would be the disarmament of the natives. . . . To disarm these people would put an end to their slave raiding . . . would prevent a great and indiscriminate slaughter of game and would be an act of authority which would at once demonstrate our supremacy."[46]

THE 1922 PROCLAMATION

By the early 1910s, Angolan gunrunners in North-Western Rhodesia were on the retreat. However, from the point of view of struggling BSAC territorial officials, the damage had already been done. Supplementing the already substantial quantities of muzzle-loaders that had entered North-Western Rhodesia in the course of the nineteenth century, the additional trade guns introduced into the country at the height of smuggling in the 1900s consolidated earlier forms of gun domestication among Africans and enhanced their potential for resisting or evading, if not colonial rule as a whole, then at least the normative apparatus that came with it. Especially in the gun-rich North-West, where the early years of colonial rule were punctuated by several episodes of gun-related violence involving both Africans and Europeans, local administrators came to the conclusion that disarmament was an absolute precondition for asserting the authority of the colonial state while curbing its African subjects' citizenship rights.[47]

In 1912, the Kasempa DC, Hazell, sought an audience with Administrator L. A. Wallace in Livingstone, the capital of the newly unified Northern Rhodesia, to address the urgent need for the immediate disarmament of the district. Both the Lunda and the Kaonde of Kasempa, he submitted, "are far from being in a proper state of control, and . . . this has been brought about . . . by the fact of their ability through the possession of firearms to defy and resist authority."[48] Wallace, however, thought that Hazell's scheme—which envisaged the detention of all chiefs "pending surrender of such guns and powder as they and their people possess" and which would require the deployment of at least four hundred police[49]—would merely multiply the chances of armed confrontation by forcing the "natives . . . to decide without any warning whether or not to obey an order for disarmament on the spot." He thus put forward a counterproposal for the more gradual registration and licensing of firearms.[50]

In April of the same year, Wallace reiterated his views in an ad hoc meeting with H. J. Gladstone, the high commissioner for South Africa, in Cape Town.[51] A comprehensive "Arms and Ammunition Proclamation" imposing a registration and licensing regime was drafted. Informed by the belief that "the inhabitants of the Kasempa and Lunda Districts have beyond question far more guns than they ought to," it gave local BSAC officials "full power of refusal of a licence"; the expectation was that "after the law comes into operation every native-owned rifle [sic] will become prima facie illegally owned unless a licence can be produced."[52] Though falling short of promoting complete African disarmament, the proclamation's obvious objective, as pointed out by the secretary for native affairs, was to "[hinder], in every legitimate way, the natives of this Territory from acquiring additional firearms and fresh supplies of ammunition."[53] However, the arrest of Sakutenuka, the district's most notorious outlaw, in May 1912 (followed by his swift execution in October), and the desire not to further antagonize local Africans meant the proclamation was cast aside, and the attempt to reestablish a modicum of order in the troubled northwestern marches of North-Western Rhodesia would now rely on the Collective Punishment Proclamation, which the colonial secretary deemed "more regular and satisfactory than any merely punitive measures."[54]

In 1914, the sale and exchange of modern rifles throughout Northern Rhodesia were made conditional upon obtaining the permission of the administration.[55] But it was only the post-WWI Treaty of St. Germain for the Control of the Traffic in Arms that brought back to the fore the still unresolved question of the former North-Western Rhodesia's missing gun legislation. After a very convoluted legal history that need not retain our attention, "The Northern Rhodesia Firearms Restriction (Natives) Proclamation 1922" was finally gazetted on 6 January 1923. Its most important provisions stipulated that "no native shall be entitled to have or possess arms or ammunition in the Territory, unless by the written permission of the Administrator" or authorized district officials.[56] Alongside "permits to possess"—the concession of which was made dependent upon payment of a fee of sixpence from the beginning of 1924—identically priced "permits to transfer" were also introduced to regulate gun exchanges.[57]

The debate that followed the issuing of the proclamation of 1922 shows that, with the disruptions of the early 1910s a full ten years behind them, local officials in the North-Western Province and elsewhere were now more willing than their predecessors had been to view firearms as more than just a threat to law and order (which they undoubtedly were) and to consider the multiple culturally specific meanings with which muskets had been endowed since their introduction in the region in the course of the nineteenth century. Practical troubles in implementing the law were anticipated in the Solwezi subdistrict, where "ten Messengers, unable to read, will have to deal" with as many as "about 4,000" Kaonde gun owners, all of whom were wont to view firearms not only as fundamental hunting tools, but also as an essential lubricant of social relationships. Ownership permits posed an especially intractable problem, since "almost every adult male native possesses one or more native guns, which are looked upon as an everyday article of barter."[58] The difficulties were much the same among the southern Kaonde, for in Kasempa "natives . . . own many thousands of muzzle-loading guns. Some natives own as many as 4 or 5: even women own guns. These guns are no less important to them as a form of currency than as weapons."[59] Hall, the Kasempa DC, thought that ample time should be given for registering the guns, since "wholesale confiscation" was initially to be avoided so as not to "seriously antagonize nearly the whole population."[60] In light of the "extraordinary" ubiquity of muzzle-loaders among the Kaonde, Hall's NC was even more pessimistic than his superior about the licensing exercise's real prospects of success. "I can imagine that an *efficient* registration system in the Kasempa district might easily present a number of problems comparable (in a lesser degree of course) to an attempt to register the ownership and transfer of sovereigns or half-crowns in the United Kingdom and that the stamping of a number on a gun would perhaps in a year's time be as helpful as the stamping of a number on a sovereign."[61] In fact, he thought the aim of rendering "the possession of guns unpopular and unprofitable" would be more easily achievable by concentrating efforts on preventing illegal importation of powder from the Congo.[62]

The same perceptive official pointed out that gun ownership in the Kasempa district was "a matter requiring greater delicacy of treatment, probably, than in other districts."[63] But misgivings were also expressed in Barotseland and Balovale. For instance, Litunga Yeta, Lewanika's successor, and a number of his councilors—long accustomed to enjoying ready access to firearms and to exploiting their symbolic power as markers of royal status and seniority—questioned the need for permits to transfer. "It would be difficult," they explained to the resident magistrate, "to obtain permits every time when one wishes to send one of his family to go and shoot game and ducks as many gun owners are old people and members of the Khotla [i.e., royal council] and cannot always go out shooting as they have to attend to official business."[64] The adoption of strategies of evasion on the part of Luvale hunters was anticipated in Balovale. Given that the majority of his messengers were illiterate, the subdistrict's NC was certain that theoretically illegal "temporary transfers" of individual permits would become very common among Luvale men: "if the owner of the gun hands over his permit to possess to the person whom he wishes to hunt for him the chances of being discovered would be slight."[65]

Despite all of these concerns, the new gun legislation was put into effect, though, as the SNA himself had recommended, it was initially applied with "a certain amount of latitude."[66] While no violent opposition to registration manifested itself, discontent in Kasempa was clearly palpable during the law's "difficult" first year.[67] It was probably not coincidental that it was in the course of 1923 that the district witnessed the first strong patrol by the Northern Rhodesia Police since the beginning

of the First World War. Following the course of the Lunga and comb-
ing the Jiwundu swamp near Solwezi, the police arrested a number of
tax defaulters and seized as many as a hundred unregistered guns.[68]
The Kaonde—who clearly feared that registration would in due course
be followed by the requisition of their most treasured possessions—reacted
by seeking to dodge the proclamation's provisions. Much evasion also
took place among the Lunda of the Balunda or Mwinilunga subdistrict, where the registration exercise began in September 1923. After
one year, permits to possess had been issued for six hundred guns, but
the DC was "pretty sure there [were] over 3,000 in that Sdt."[69]

Prosecutions and confiscations for failure to register and obtain
the necessary licenses began in earnest in the Kasempa district in the
summer of 1924.[70] This had the unintended immediate effect of bringing registration to a complete halt. Since "everyone found with an unregistered muzzle loader got two month [sic] IHL [Imprisonment with
Hard Labour] without the option of a fine and the gun was confiscated
as well," the "natives had no choice but to conceal unregistered guns."[71]
After Hall's successor reverted to "a more moderate course from the
beginning of 1925, 669 new guns were registered" and licensed.[72] To
be sure, the end of the exercise was still not in sight. Yet, the new DC
commented, this was to be expected, given that "Natives look upon
these guns as a form of currency, and up to the year 1923 were allowed
to possess them without formality. Since it is probably the ambition
of every male native to own one or more of these guns, it is fairly certain that many thousands of them must now exist in N.W. Rhodesia."
In "view of the deep suspicion with which most natives would at first
regard any law which sought to control anything they highly prized,"
the DC thought the "Firearms Proclamation ha[d] been carried out
as efficiently as could be expected," and he was confident that, "provided no harsh or repressive measures are now adopted . . . a fairly
complete registration will in due course be effected."[73] In the end, the
Kaonde and their neighbors bent the knee, though dissatisfaction with
the newly introduced licenses endured and must have underlain the
numerous manifestations of hostility to colonial taxation registered by
missionaries in 1924–1925.[74]

Slowly but surely, then, the hitherto unregulated right to possess
guns was taken away from the peoples to whom it had mattered most.
The effects of the Proclamation of 1922 were quickly compounded by
its logical sequel: the extension to the bulk of the former North-Western

Rhodesia of such game laws as had regulated African hunting in the eastern part of Northern Rhodesia since the beginning of the century.[75] The phased imposition of "native hunting licences," of course, constituted another major external interference in the life of communities whose deep-rooted hunting traditions had been strengthened by the intensification of contacts with the Atlantic economy and the widespread adoption of firearms over the course of the previous century. The colonial assault on unrestricted hunting and gun ownership in the former North-Western Rhodesia was a long time in the making, but when it did materialize, its consequences were momentous, for it marked the beginning of the end of a system of socioeconomic and cultural relationships that had dominated large stretches of northwestern Zambia for several decades.[76]

COLLABORATION AND GUNS IN SOUTHERN KATANGA

Continuing user demand and the resilience of Ovimbundu and Luso-African operators were the key factors in ensuring the availability of firearms to Africans in the still largely unoccupied northwestern part of colonial Zambia between the end of the nineteenth and the beginning of the twentieth century. In the Congo Free State, a similar situation obtained on the upper Lualaba and Lubudi Rivers—where a lively traffic in precision weapons existed between Mambari caravaneers and Force Publique mutineers[77]—and along the upper Kasai. In the latter area, the bulk of Chokwe weapons—Liège-made lazarinas, highly prized on account of their lightness, maneuverability, and technological accessibility—continued to be provided by Angolan gunrunners well into the early 1910s.[78] Southern Katanga, on the other hand, witnessed a more sustained military and policing effort on the part of the CFS, and Angolan involvement in the economy of the region was correspondingly shorter-lived. The presence of Ovimbundu caravans around the late Msiri's old capital, Bunkeya, was reported in 1894–1895.[79] One such caravan, directed to Ntenke's, was said to be five hundred strong and to be carrying both guns and gunpowder.[80] Already in 1896, however, the "old trade roads" to the Yeke of Garenganze were said to have been "shut"; as a result, people were short of calico.[81]

With Angolan merchants increasingly marginalized, continuing Yeke access to firearms came to depend less on the survival of precolonial commercial networks than on the maintenance of a solid alliance with the newly arrived Europeans. Late in 1893, Legat, the first Force Publique officer in charge of Lofoi, the CFS's military post to the east of Bunkeya, was replaced by Clément Brasseur, a former sublieutenant in the 3ème Chasseurs à Pied and a true villain of the piece if there ever was one.[82] Having added "some fifty" soldiers to his predecessor's limited force, Brasseur—who regarded "all the niggers as first-class Jews . . . obeying force only"—set about to "thoroughly take in hand" the country.[83] Brasseur was determined to maximize the state's—and, therefore, his own—revenue in ivory.[84] With this aim in mind, he inaugurated a reign of terror that left a vivid impression on both local Africans—by whom he was nicknamed "*Nkulukulu*," after a bird "whose inner wings are bloody red. Now, the natives say, Mr. Brasseur was only happy when he had blood up to his armpits. Then he looked like the bird in question"[85]—and Plymouth Brethren missionaries, whose protests he warded off with the threat of imprisonment and whom he eventually forced temporarily to abandon the Lofoi area late in 1894.[86] Brasseur, wrote Crawford, "rul[ed] with a far higher hand than ever Msidi did."[87] His extractive methods were especially crude. Replicating, and modeled on, the Yekes' earlier pillaging economy, they revolved around the deployment of so-called sentries— unsupervised Force Publique soldiers stationed in prominent villages and entrusted with the task of collecting "all the ivory in the country, and see that the Chiefs and the hunters sold none elsewhere"[88]—and the staging of large-scale military expeditions intended to impress upon local leaders the need to comply with instructions relating to taxation in kind and labor.

The recasting of Mukanda Bantu's followers, now settled at Lutipisha, in the vicinity of Lofoi, into a rapid deployment force at the service of the state was a key part of Brasseur's plans for Katanga. His strategy worked because it was entirely consistent with the Yekes' own interests and, specifically, their resolve to overcome the near-fatal crisis into which the Sanga rebellion and the death of Msiri had plunged their polity and society. The alliance with the CFS offered Yeke men of violence the chance to roll back the years and check the assertiveness of the Sanga and other former subjects—those same subjects whose new-found courage the Yeke belittled in song: "You were poor," a Yeke tells an imaginary antagonist. "You even boast in the country. . . . That's because the Belgians are in Bunkeya."[89] In the process, the Yeke military played a critical role in shaping from below a violent system of oppression which the Europeans coordinated and benefited from, but whose ultimate meanings and direction they did not always control or even fully understand.

Beginning in 1894, Mukanda Bantu and his gunmen participated in all of the major expeditions carried out by Brasseur and/or his subordinates. In July of that year, about one hundred Yeke joined Brasseur's foray against Mutwila, the former Sanga rebel chief (see chapter 3). The fortified villages of Mutwila and his brother Mukemwa ("Moéma") were stormed by the CFS-Yeke force.[90] Numerous children and women were killed during the "great slaughter" that followed the attack,[91] the Yeke behaving—in Brasseur's own words—as a "pack of famished wolves scrambling for a corpse."[92] Brasseur, his soldiers and auxiliaries returned home with sixty female prisoners, fourteen ivory tusks, some guns, and four kegs of gunpowder.[93] Mukanda Bantu was naturally much impressed by the demise of the old enemy and insisted on escorting Brasseur back to Lofoi with two hundred of his people.[94] In May 1895, the followers of the then Kitobo ("Chitoba")—a Nwenshi or Bena Mitumba leader—revolted against Kalasa, who had been selected to replace the deceased Yeke mwanangwa, Kalala. Brasseur and his recently arrived Hausa soldiers sided with the Yeke and, accompanied by 150 fighters of the by now "very devoted" (très dévoué) Mukanda Bantu, proceeded against the rebels and scattered them.[95]

The following year, the auxiliaries of Mukanda Bantu—whom Brasseur had taken to referring to as his "commander in chief"[96]—accompanied the Belgian official on his longest tour yet. Targeting another historical enemy of the Yeke, the Luba of the Upemba Depression, this expedition was described by the missionary Campbell as possibly the "the biggest and most bloody" of Brasseur's journeys.[97] By this time, the bond between the Yeke warrior king and Brasseur was such that the latter felt confident enough to send the former and one of his war parties on a solo attack against "Kasangula, a Bona-Mitumba chief, who was known . . . to have ivory." Campbell saw the Yeke return with prisoners, "twenty-one heads and the same number of tusks of ivory."[98] Since such a raid was an almost exact replica of one of Msiri's former slave forays, it is perhaps not surprising that "one of the male prisoners . . . was sacrificed at the grave of old Msiri unknown to the State."[99] Indeed, so closely intertwined had Brasseur's Force Publique and his Yeke auxiliaries become that many war traditions of the latter—including mutilations and the ritual purification of severed enemy heads—were routinely brought into play during campaigns which were ostensibly led by the former, but which were also clearly informed by Yeke motives and interests. Writing some years after the events, Judge Rutten, a future governor general, admitted that "barbarous acts" (actes de barbarie) had been committed during Brasseur's expeditions. He then explained that this was the result of his "being always accompanied by numerous Yeke auxiliaries."[100] Brasseur, Rutten surmised, had perhaps been obliged to tolerate the "customs" (coutume) of his "most loyal" local allies (ses plus fidèles auxiliaires).[101]

Brasseur finally had his comeuppance in November 1897, when he was fatally wounded during an attack against the fortified settlement of the Swahili trader Chiwala, on the upper Luapula River. Despite missionary expectations to the effect that his death might bring about "a more humane state of affairs in Katanga,"[102] widespread CFS-Yeke predation remained rampant in the late 1890s, the height of the short-lived rubber regime in southern Katanga, while the Sanga and other autochthonous groups continued to live in dread of both the CFS's sentries and Yeke militias.[103] The last major Yeke contribution to the colonial conquest of Katanga took place in 1899. Fittingly, from their point of view, it consisted of an expedition against another former Sanga rebel commander, Mulowanyama. After killing the Belgian Fromont, Mulowanyama and his people sought shelter in some caves near their village. Besieged by the attackers, they held out for three months, "only a handful yielding, and then the Belgians found all were dead, and they recovered a large number of guns from the decomposing bodies. It was a horrible thing thus to go into their own grave."[104]

In theory, following the Acte Général of the Brussels Conference of July 1890 and related royal decrees, the only firearms that private traders could legally import into the CFS and sell to Africans were "smoothbore flintlocks" (fusils à silex non rayés), as well as "trade gunpowder" (poudres communes dites de traite).[105] In practice, as the Belgian government that took over Leopold's private colony in 1908 knew only too well, CFS's officials had been wont to both sell and freely distribute not only flintlocks, but also "a large number of percussion-lock guns" (un grand nombre de fusils à piston) to the many local allies on whom they depended for the "pacification" of the huge region they had been entrusted with.[106] Like other African associates of the CFS, the Yeke benefited a great deal from such early instances of state patronage.[107] In July 1895, having escorted Brasseur and his soldiers on a tour to the west and south of the Lofoi station, Mukanda Bantu and his fighters were rewarded with a present of powder and caps.[108] One year later, Brasseur offered the same recompense to the Yeke who had accompanied him to Lubaland.[109] Campbell described this donation as "the State quota of powder and percussion caps."[110] As we have seen in chapter 3, the Yeke had adopted percussion-lock muskets on a significant scale during the last years of Msiri's reign. Their alliance with the CFS throughout the 1890s and beyond meant that colonialism did not cut short this process of technological engagement, but rather gave it new impetus—as attested, perhaps, by the fact that in 1905 more than seven hundred free and exceptional permits to possess "*armes de guerre*" were issued to African leaders by the Comité Spécial du Katanga.[111]

Supported by the CFS, and "armed with State guns, caps and powder,"[112] the Yeke were able to rebuild their regional ascendancy. Early in 1892, in the immediate aftermath of Msiri's death, Mukanda Bantu had cut a powerless figure, surrounded only by "a handful of people."[113] Four years later, however, Lutipisha, the Yeke temporary capital near the CFS station of Lofoi (Lukafu from 1899), was already being described as "seething" and second only to Mwansabombwe, the great eastern Lunda capital on the lower Luapula River.[114] At the end of the same year of 1896—when the Plymouth Brethren reestablished themselves in the Lofoi area—Lutipisha's population was estimated at four thousand.[115] The figure of five thousand was given a few months later by Campbell,[116] even though, by c. 1900, the same observer had downsized his estimate to (a still considerable) three thousand. Of these, only one-fifth were reported to be men.[117] The large number of women in Mukanda Bantu's headquarters was also noticed by Lily George, a newly arrived female missionary.[118] When juxtaposed with the evidence presented above, Lutipisha's skewed demographic composition in the late 1890s is an almost certain indication that Yeke fighters had successfully used their participation in state-promoted razzias to rebuild their stocks of slave wives and field hands. Writing in 1905, Judge Rutten was honest enough to acknowledge that "the surroundings of Lofoi and Lukafu were inhabited by women formerly captured in war."[119] Naturally, besides the Yeke, the Force Publique and the Lofoi station also benefited from the captive labor thus obtained.[120] Early in 1895, for instance, Brasseur openly admitted to having given "a woman each" to some newly recruited soldiers.[121] Altogether, until the early 1900s, behavior towards indigenous women was, as Jenniges put it, "very arbitrary" (fort arbitraire).[122]

During the violent 1890s, then, the imperatives of surviving the consequences of the Sanga rebellion and reconstituting their hegemony in tandem with the CFS dominated the strategy of the Yeke. In this context, such firearms as they continued to have access to as a result of state patronage were still primarily deployed for military purposes. The process of "normalization" that followed the abolition of the sentry system in the early 1900s and, especially, the Belgian takeover of 1908 rendered such patterns of gun usage obsolete. Equally important were the coeval effects of the inception of industrial mining in the erstwhile southern periphery of Msiri's warlord state. By 1910, the year of Mukanda Bantu's death and of the enforced return of his followers to Bunkeya, Yeke commoners were already being drawn into the orbit of such mines as Kambove and the Étoile du Congo or of the itinerant traders who kept the same mines supplied with foodstuffs.[123]

The Yeke who did not travel to the mines or sell them the produce of their agricultural labor were still faced with the problem of meeting their fiscal obligations towards the state. Large-scale hunting offered them a solution.[124] Up to 1910, the year in which a new game law was passed, only elephant hunting was strictly regulated in the Congo. Even after that date, "collective"—as opposed to individual—hunting permits were still expected to be issued free of charge to Africans.[125] This comparatively lenient hunting legislation explains why all of my Yeke informants described the colonial period as an era of unrestricted game hunting. Game laws aside, the transition to commercial hunting also owed something to the animal wealth that (notwithstanding Ivens's pessimistic predictions in the 1880s) seems to have characterized Bunkeya and surrounding districts in both the nineteenth and twentieth century. In 1895, game of every sort abounded in the Lufira Valley and the Kundelungu Range; as late as 1897, Brasseur remarked that Katanga still "teemed with elephants" (pullule d'éléphants).[126] In that same year, "native hunters" brought the missionaries at Mwena, near Lofoi, "more game than [they could] use."[127] Also important, of course, was the fact that large-scale hunting was consistent with precolonial Yeke economic pursuits and sociocultural predispositions, for—as has been seen in chapter 3—organized hunting had been part and parcel of the political economy of Garenganze. Mukanda Bantu himself was a passionate hunter, being often found "at his old work of roaming the woods for venison."[128] The growth of commercial hunting in southern Katanga during the first two decades of the twentieth century was so remarkable that Campbell saw fit to comment on it in his published recollections. Comparing the situation of the Congo with that which obtained in the almost fully disarmed North-Eastern Rhodesia, the long-serving missionary could not help noticing that, thanks to the large number of flintlocks and percussion-lock muskets available to them, Africans in the Belgian colony "hunt and kill more game to-day than in the early days. In the Katanga many natives do a regular trade in game meat at the mines and in the towns along the railway."[129]

By 1932, the first year for which comprehensive local data are available, recognizably Yeke chieftainships housed some 450 licensed and an estimated 150 unlicensed (mainly English and American) percussion muskets—about one-quarter of all of the Western-made guns then circulating in the Territoire de Jadotville.[130] To such weapons we must add the increasing number of (presumably unlicensed and unregistered) percussion-lock muzzle-loaders (poupous) which began to be manufactured at about this time using gun scraps, recycled industrial parts, and homemade pieces (see figures 0.1 and 4.2). Such technological innovation was the product of the interplay between high user demand and the refinement of a set of preexisting technical competences. It represented a considerable achievement in the eyes of Yeke hunters, for the names are still presently remembered of some of the first local gunmakers who died in the 1940s or 1950s.[131] The spread of homemade hunting guns went hand in hand with the development of techniques for the local production of gunpowder—techniques in which, as we know, the Yeke had not invested at the height of their involvement in the Atlantic economy. Though most of the trade gunpowder employed by Yeke hunters came from Jadotville (present-day Likasi) and other industrial and market towns, local manufacturers could now produce a serviceable powder by mixing charcoal and imported "soda" (presumably, saltpeter).[132]

Early twentieth-century Yeke hunting practices built on precolonial precedents, but differed from them in some crucial respects. In the new circumstances, the closed and centrally controlled hunting guilds that Msiri had once patronized were sidelined in favor of more individualistic forms of hunting.[133] Although the Yeke kings' permission was still required to poach (theoretically) state-protected elephants, most other types of hunting and the guns on which they depended were now decoupled from royal or state power. In a socioeconomic context characterized by what Belgian officials considered to be urban-derived

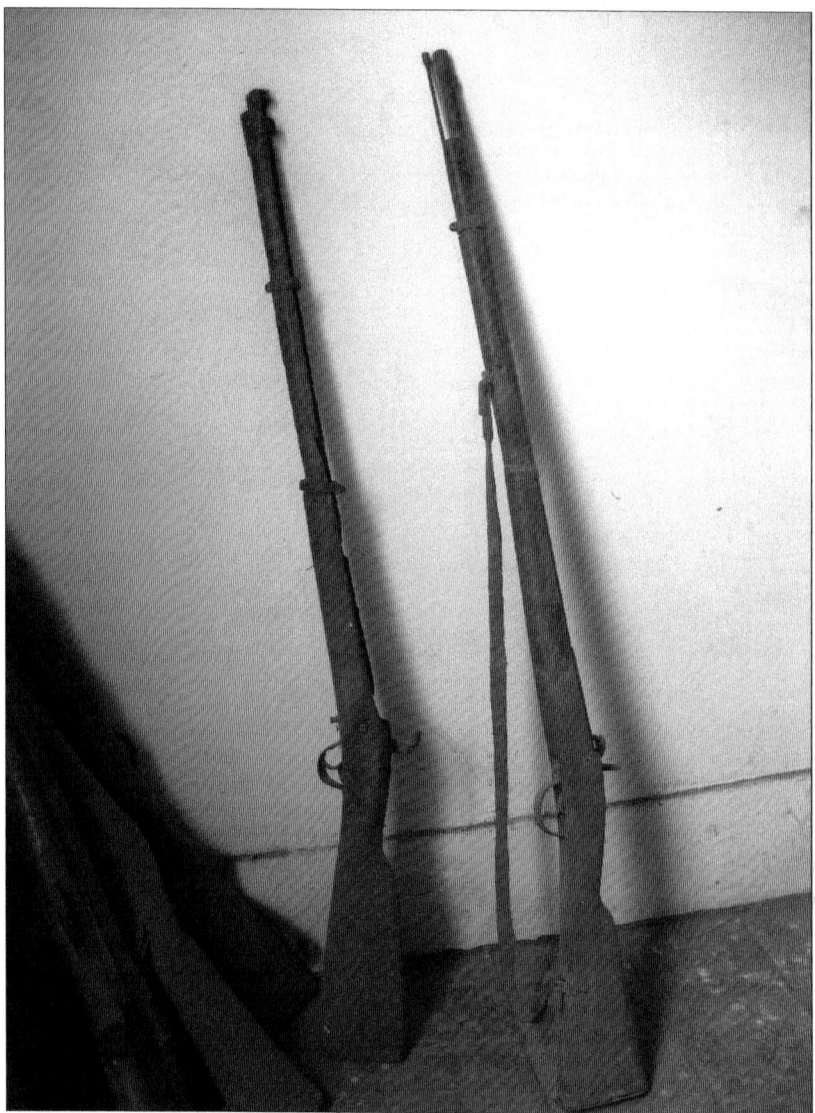

FIGURE 4.2 Two homemade poupous, Bunkeya, DR Congo. Photograph by the author, August 2011. According to Belgian expert Paul Dubrunfaut (personal communication to author, 30 June 2012), the first percussion musket from the right displays a high degree of technical proficiency. It was manufactured using gun scraps and locally made pieces. The upper barrel band might have come from an original Enfield rifle, while the only partly visible lock could be a local imitation of a French back-action lock. In Dubrunfaut's estimation, this piece is likely to be between fifty and a hundred years old. The second gun from the right, without barrel, would have been an almost identical specimen, probably originating from the same workshop.

"*habitudes de liberté*,"[134] percussion muskets, whether locally made or imported, acquired a function similar to that which the lazarinas had fulfilled among the Luvale and their neighbors before the imposition of rigid game and gun control laws by the Northern Rhodesian administration in the early 1920s. Losing their earlier significance as centrally monopolized means of military and economic domination, guns were now recast as individually owned tools of production and emblems of male identity.

In twentieth-century southern Katanga, hunting was construed as a more prestigious activity than trapping.[135] Partly, this was because the poupous available to most Africans magnified the dangers of the hunt. As a very widespread song (here given in translation from a chiBemba version) pointed out, they misfired frequently: "You, my gun, you shamed me / In this bush . . . there are animals, father / There are animals, my gun, you shamed me." The "shame" (honte) thus brought upon the hunter was likened to failing to have sexual intercourse with a woman. The song, indeed, is often said to have a double meaning, applying equally to hunting pursuits and conjugal life.[136] A different version urged the gun—the father of the previous song—not to humiliate the hunter, but rather to impress him with its strength ("*montre-moi ta force, mon fusil*").[137]

↩

While the upper-Zambezi gun societies withered in the aftermath of the introduction of new gun and game laws in the 1920s, the Yeke gun society, largely because of the mode of its original incorporation into colonial structures, retained vitality. From the 1900s, guns were de-coupled from state and military power in southern Katanga, but their large-scale deployment in commercial hunting and the emergence of a local gun industry meant that they preserved a more central economic and cultural position among the Yeke than they did among neighboring Northern Rhodesian communities. And the Congo's troubled twentieth-century history ensured that the story of the gun among the Yeke did have a sting in its tail. The Yekes' continuing familiarity with firearms throughout the colonial period and the assertive masculinity of which it was both a cause and a consequence are likely to have played a role during the secession of Katanga, partly accounting for the enthusiasm with which the Yeke volunteered for service in the Gendarmerie of Interior Minister, and future Mwami, Godefroid Munongo.[138]

During the secession, the relationship between firearm technology and Yeke society was once more reforged, as the gendarmes' rifles and homemade weapons were reinvested with a predominantly military meaning. Msiri and Mukanda Bantu would have felt at home in early-1960s Bunkeya.

PART III

Resisting Guns in

Eastern Zambia and Malawi

5 ⮑ "They Disdain Firearms"

The Relationship between Guns and the Ngoni

THE PREVIOUS three chapters of this book may have given the impression that the appeal of firearms was universal across the interior of central Africa and that only comparative commercial isolation prevented specific, small-scale, and/or marginal groups, such as the Ila of the Kafue River, from taking full advantage of the new technology. This would be a profoundly mistaken conclusion. The reception of foreign artifacts—this book has argued throughout—cannot be divorced from preexisting social structures, political circumstances, and cultural values. It is these latter forces that dictate the shape and direction of local processes of domestication—processes though which technological imports are redefined and infused with both material and symbolic functions that do not necessarily replicate those for which the objects in question had first been devised in the contexts of their original production. This chapter focuses on the Ngoni of present-day eastern Zambia and Malawi in the nineteenth century to foreground the simple—yet crucial—point that the *rejection* of a given exogenous technology is no less socioculturally motivated than its adoption. All the available evidence shows that the Ngoni did not deploy firearms in warfare. This—I will argue—was a deliberate choice, rather than the enforced consequence of economic insularity or the technical deficiencies of the imported hardware of violence available to them.

The first part of the chapter examines some salient episodes in the precolonial histories of the principal Jere and Maseko Ngoni groups with a view to isolating the key features of their relationship with firearms. Departing from dominant scholarly depictions of Ngoni society and economy, my analysis shows that, despite having repeatedly experienced the military potential of the new technology and having had ample opportunities to gain access to it through trade, both Jere and Maseko Ngoni fighters consistently resisted its adoption for war purposes. An explanation for this only apparently aberrant behavior is offered in the second part of the chapter, which teases out the logic behind the Ngonis' military conservatism. Their technological disengagement is here understood less as a "failure" to adjust to changed circumstances than in terms of the sociocultural forms that gave rise to it. The Ngoni, this chapter will contend, objected to fighting with firearms because they appraised this particular technological import in pejorative terms, viewing it as posing a threat to dominant notions of masculinity and honor constructed around close combat. By making hand-to-hand combat obsolete, firearms also threatened to stymie the progress of youthful fighters of captive origins through the ranks of the Ngoni military. In this context, the new technology was doomed from the start. As will be argued in the next chapter, it took such revolutionary events as the violent imposition of colonial control and the subsequent demise of the Ngoni's war-centered social system to fundamentally transform their evaluation of firearms and terms of engagement with them.

EXPERIENCING FIREARMS

Early in the second half of the nineteenth century—following a four-decade-long process of migration, military conquest, demographic expansion, and political segmentation—a number of Ngoni groups acquired a dominant position over large swathes of the territory corresponding to present-day Zambia and Malawi. Paramount among these were the Jere Ngoni, the motley descendants of war leader Zwangendaba Jere, one of the protagonists of the Mfecane (see chapter 1). Following the defeat of his ally, Zwide of the Ndwandwe, at the hands of Shaka's Zulu in the late 1810s, Zwangendaba had led his growing (and growingly heterogeneous) followers as far north as Ufipa, in southwestern Tanzania.[1] Upon Zwangendaba's death in about 1845, his "snowball state" (to use John Barnes's famous characterization[2]) fragmented into several autonomous sections, the most important of which were headed by two of his sons: Mpezeni (b. c. 1830[3]) and M'Mbelwa (b. c. 1840[4]). By the 1860s, the former and his partisans were temporarily settled among the

MAP 5.1 Eastern Zambia and Malawi in the late nineteenth century.

Nsenga of present-day Petauke district of eastern Zambia and Mara-via district of western Mozambique, and would shortly thereafter shift the heartland of their polity eastward, overrunning the Chewa of the *Mkanda* and others, near present-day Chipata. In the meantime, the Jere Ngoni who would eventually recognize the supremacy of young M'Mbelwa had succeeded in imposing their sway over the Tumbuka and related peoples of Mzimba district, northern Malawi, in the course of the 1850s.

Equally meteoric was the rise of a separate "Ndwandwe offshoot," who crossed the Zambezi independently of Zwangendaba in the late 1830s and became known as the Maseko Ngoni.[5] In about 1870, after spending some years to the east of Lake Malawi and having been defeated by yet another Jere splinter group—Zulu Gama's Ngoni—the Maseko Ngoni, led by the regent Chidiaonga, retraced their steps to the southwestern corner of Lake Malawi, incorporating into their emergent polity most of the inhabitants of present-day Dedza and Ncheu districts of central Malawi. Following the death of Chidiaonga in the early 1870s, Chikusi, the son of the late Maseko paramount, Mputa, took over the leadership of the group, though his ascendancy would soon be challenged by his paternal cousin Chifisi, Chidiaonga's own son.[6]

All of these migrant groups—notwithstanding the recasting of their southern identities that the assimilation of large numbers of captives was bringing about—claimed origins in South Africa, spoke northern Nguni dialects (which, however, were being rapidly marginalized to the advantage of the local languages spoken by captive wives[7]), and were deliberately organized and trained for war, based on the principle of age-set regiments and the (theoretical) military mobilization of all able-bodied men. The Jere Ngoni of M'Mbelwa I and Mpezeni I, and even the Maseko Ngoni of Chikusi, have commonly been portrayed as "formidable" military machines that attached "little importance" to external trade, the captives and cattle necessary to energize their political economies being obtained mainly by means of "frequent raids" against their neighbors.[8] As already noted by Leroy Vail,[9] modern scholarly depictions owe something to early missionary characterizations: for W. A. Elmslie—at Njuyu, the first Christian mission station in the northern Ngoni's heartland, since 1885—"cattle-lifting was a constant occupation in the dry season" for the "proud warriors of Ngoniland," whose "hordes" kept surrounding districts in a permanent "state of terror and distress."[10] Without wishing entirely to deny all historicity to these descriptions (for all three Ngoni groups did maintain a degree of contested military ascendancy until the arrival of European forces in their respective areas at the close of the nineteenth century), the three subsections that follow contend that both the Ngoni's military might and their putative economic insularity have been hitherto somewhat overestimated.[11]

Partly as a result of the violence precipitated by the advancing frontier of the long-distance slave and ivory trades, defenses around settlements became more and more common in the central savanna in the second half of the nineteenth century.[12] When fired from within heavy stockades, flintlock and percussion-lock muzzle-loaders, their limited range and long reloading time notwithstanding, could prove deadly enough to deter even the most determined of assailants. Indeed, an awareness of the dangers posed by muskets began to develop among the northern Jere Ngoni from as early as the 1860s, when a raiding party that included the newly crowned M'Mbelwa was repulsed by the then Mwase Kasungu, an important Chewa leader who had recently obtained some guns from the Arab-Swahili settlement at Nkhota Kota (see chapter 1).[13] The defeat suffered on this occasion was still vividly remembered in Mzimba district several decades after its occurrence. The Ngoni attackers—the missionary Cullen Young recorded—had been forced to "[spend] the hottest hours of a very hot day in October lying flat on their faces among the ash and cinder of the burnt plain while the Chewa bullets hummed and droned above their heads."[14] The few casualties sustained outside Mwase Kasungu's fortified capital were apparently sufficient to persuade the northern Ngoni leadership never again to allow M'Mbelwa to take part in military expeditions, and, more importantly, to strike an alliance with the same Chewa ruler, who from the 1870s agreed to lend "frequent" "armed assistance" to M'Mbelwa in exchange for being spared further attacks.[15] Mwase Kasungu was one of the very few Chewa chiefs of central Malawi to retain a large degree of independence vis-à-vis the Ngoni of Mzimba. There is little doubt that his success owed much to the comparative strength of his firepower and fortifications, and to the respect with which the Ngoni regarded both. Mwase Kasungu is said to have made "effective use of the *linga* system of defence, which involved the alternative use of gunners and archers inside the high walls of a stockade."[16] Because of his ability to thwart the Ngoni, his territory attracted numerous refugees from surrounding Chewa-speaking areas.[17]

Further setbacks were experienced by the northern Ngoni's military apparatus in the 1870s and 1880s. More often than not, their successful antagonists had access to firearms, though, of course, imported guns need not always have been the determining military factor. Around 1870, a few Arab-Swahili musketeers took part in the decisive battle between a large coalition of Bemba chiefs and the followers of Mperembe, a brother of M'Mbelwa I and Mpezeni I, who had spent much of the 1850s and 1860s raiding the southern borders of Bemba country and had then relocated to the south of Lake Tanganyika. The defeat they endured convinced Mperembe's people to take up residence with M'Mbelwa, in northern Malawi.[18] If "it is not clear how much the traders' guns contributed" to the Bemba's victory against Mperembe's Ngoni,[19] the Arab-Swahili presence among the Senga of the upper Luangwa River beginning in the late 1870s—and the numerous muskets they made available to them in exchange for access to their rich ivory reserves—had certainly more than a little to do with this much-victimized group being finally able to stand up to northern Ngoni raiding parties.[20] Once more, having come up against well-armed opposition, M'Mbelwa's Ngoni resolved to enter into an agreement with the emerging power, limiting their raids against the Senga in return for some "annual gifts of ivory."[21] The military alliance was put to the test in 1887, when a Bemba foray against the Senga of the *Chifunda* was successfully repelled by a combined Senga-Ngoni-Bisa force.[22] Ng'onomo, M'Mbelwa I's senior military commander, is said to have returned home with the conviction that "if it had not been for these [Senga] with their guns helping them, none of the Angoni would have returned from the Babemba."[23] In this instance, then, the use of heavily armed foreign partners stood the northern Ngoni in good stead. In 1882, however, not even Mwase Kasungu's guns had proved sufficient to prevent "the sharp reverse suffered by an *impi* . . . at Nkhota Kota," whose ruler, the Jumbe (that is, the representative of the Sultan of Zanzibar), "had accumulated up to 2,000 firearms."[24]

Particularly significant for the internal balance of M'Mbelwa's kingdom in the 1880s was the resistance put up by the Lakeside Tonga. Both imperfectly assimilated captives and separate subordinate leaders rebelled against Ngoni exactions in 1877 and holed themselves up in impregnable redoubts along the lakeshore.[25] The capital of the Tonga leader Mankhambira, at Chinteche, was encircled by "a double, and at some places a triple stockade, several miles in length," and provided shelter to "more than 1,000 huts."[26] Of similarly "enormous strength" was Marenga's "triple stockade" near Bandawe.[27] A great victory was scored by the rebels in late 1877–early 1878, when, following the mass desertion of a contingent of Tonga captives from the Ngoni district of Elangeni, the pursuing army was put to flight by Mankhambira's fighters in what would become known as the "battle of the Chinteche Stream."[28] The Ngoni historian Yesaya Chibambo and others have attributed the Tongas' success to their knowledge of local terrain and subsequent ability to entrap the besieging Ngoni.[29] But a coeval description of the confrontation points instead to the (presumably still few) guns that the Tonga of Mankhambira had recently managed to obtain as the decisive factor. The attackers, a Ngoni eyewitness told the missionary Robert Laws, "had come to the stockade on the N. side and tried to get in, some by climbing over the fence, some by going round each end of it. They saw some men peeping out, they heard guns fire, and then saw their men tumbling down from the fence or at its ends. This they thought too hot work, & turning fled. Mankambira's people pursued them for several miles, till the edge of the plain was reached; then the Mangone made another stand, but again the guns proved too much for them and they fled to the hills."[30]

From the end of the 1870s, thanks to enhanced contacts with the Arab-Swahili traders installed on Lake Malawi's eastern shore and at Nkhota Kota, a process of accelerated Tonga rearmament was under way. It is certain that Marenga, Mankhambira's Tonga rival, paid a visit to Nkhota Kota in the latter part of 1878. At the beginning of the following year, a "party of Arabs" was reported at Kang'oma's, while the neighboring Mankhambira was rumored to be resenting the Livingstonia missionaries for hampering "his trade of kidnapping slaves."[31] The connection between the sale of slaves, the acquisition of guns, and Tonga defensive requirements was made explicit by the Free Church of Scotland missionaries late in 1881, when the Bandawe Mission Journal remarked that the firearms that Mankhambira and Kang'oma were obtaining in exchange for people were being used to keep the still threatening Ngoni at bay.[32] At the beginning of 1885, acting British consul Goodrich described Chinteche, Mankhambira's headquarters, as a "well known centre of the slave trade."[33] By this time, firearms had spread widely among the Tonga,[34] increasing their chances of responding aggressively to continuing Ngoni forays along the lakeshore and generating much disquiet among the latter. At some point in the mid-1880s, the missionary Elmslie witnessed (or heard of) a group of Ngoni raiders being put to flight after being surprised by their heavily armed Tonga pursuers.[35] In 1887, the aforementioned Ng'onomo was said to have refused to attack the now gun-rich Tonga of Chinteche in the absence of support from Senga musketeers.[36]

The almost complete absence of firearms in Mzimba district—one which enabled European visitors in the late 1870s regularly to stage highly symbolic demonstrations of the deadliness of the technology at their disposal[37]—was thus not a consequence of the northern Ngoni lacking the wherewithal to assess the military efficacy of guns. Nor, I am presently going to show, was it the result of their shunning all contacts with potential importers. Indeed, there are enough indications to suggest that, notwithstanding the common historiographical stereotype, the Ngoni of M'Mbelwa I and his subordinate chiefs and army leaders were not entirely excluded from long-distance trading networks, though their involvement in the Indian Ocean economy seems only occasionally to have translated into a willingness to exchange captives, whom they rather tended "to incorporate . . . into their own tribes."[38] Some two years before Laws's first visit to the northern Ngoni in the autumn of 1878, an "Arab from Mataka's," a Yao center on the eastern shore of the Lake, had bought both "ivory and slaves" from Chiputula, the westernmost Ngoni territorial leader in Mzimba.[39] More information on the doings of "Arabs" or, more commonly, their "agents" was gathered by the artisans entrusted with the running of Kaningina—a short-lived outstation of Livingstonia in the "no-man's land" between the highlands Ngoni and the Tonga[40]—in the course of 1879. On 7 February, McFadyen reported the presence of some traders near his station. They were returning from M'Mbelwa "with about 30 tusks." One week later, a new party of "Arab agents passed en route for the Angoni with 8 trusses of calico." The passage of eight more traders was recorded on 1 April. They had seemingly been invited to the highlands by Mtwalo, another brother of M'Mbelwa, with a view to patching up relationships after an earlier attack on some of their colleagues. A few weeks later, it was the turn of the Swahili of Nkhota Kota to visit M'Mbelwa and purchase "a large quantity of ivory."[41] In May, a suspected "Arab" slave dealer "from Kilwa" was confronted near Kaningina by the African evangelist Fred Zarakuti, who confiscated his goods. These consisted of "4 pieces American sheeting, 1 do. black, 2 do. coarse check, 2 do. handkerchief calico and one gun."[42]

M'Mbelwa's openness to foreign trade may have been only temporary. Yet it is not surprising that such significant commercial links as existed in the late 1870s and, perhaps, beyond resulted in a few guns finally making their appearance on the Ngoni-dominated plateau. The *truly* surprising aspect is that the new technology—one that, as has been seen above, some of M'Mbelwa's allies had enthusiastically appropriated—was never deployed for military purposes by the northern Ngoni, who continued to use "no weapons but the shield and assegai."[43] By the end of 1887, the firing of muskets had seemingly become part of M'Mbelwa's court ceremonial.[44] The king himself enjoyed firing off guns "for fun," and he was frequently "beg[ging] fowling piece cartridges" from the missionaries.[45] But these initial experiments were not the prelude to a deeper engagement with the new technology. As late as 1892, the members of one of the last Ngoni impis ever to raid northern Malawi were still only armed with spears.[46]

My contention is not that foreign trade was at any point fundamental to the political economy of M'Mbelwa's kingdom. There—unlike in more settled savanna polities—the redistribution of Western commodities was never a key means to consolidate links of loyalty and subordination. The point of the above discussion is more simply to argue that if the northern Ngoni leadership had wanted to gain widespread access to firearms with a view to exploiting their military potentialities, it would have been in a position to do so. An explanation for this manifest technological conservatism in the military sphere needs to take into consideration some key principles of Ngoni social organization and their cultural manifestations. Before doing this, however, it is worth asking whether a similar diffidence towards firearms was also exhibited by the two other Ngoni communities that form the subject of this chapter.

Maseko Ngoni

When Laws visited the Dedza Highlands in the summer of 1878, the Maseko Ngoni had still had little contact with long-distance traders. Cotton clothes were mainly of local manufacture,[47] and imported goods were scarcely in evidence in Chikusi's capital. Unsurprisingly, Chikusi's Ngoni were lightly armed, with "shields, spears, and tomahawks" being their common weapons.[48] The Snider-Enfield breech-loading rifles in the possession of Laws's party generated much curiosity among the Maseko.[49] A similar reaction had been recorded by Edward Young in 1876, when he had met a three-hundred-strong Maseko raiding party in the Shire Valley. His "guns, revolver, etc." were the items by which the impi's military commander was most readily surprised.[50]

By the mid-1880s, the Maseko economic isolation was already a thing of the past, with strong trading links having been established with both Portuguese Zambezia—Walter Kerr met some Ngoni who "had brought ivory" for sale in Tete[51]—and "east coast traders," some of whom, the same explorer reported after reaching Chikusi's heartland in 1884, were apparently stationed "constantly in this district."[52] It was probably these same traders who provided the "good supply" of calico observed by Kerr among the members of the Maseko "royal circle."[53] Unlike M'Mbelwa, Chikusi, whose territory was even less richly endowed than that of the northern Ngoni,[54] was not opposed to exporting slaves, though Kerr, unaware of the extent to which captives were commonly incorporated into Ngoni societies, was probably exaggerating in describing the "district under the sway of Chikuse [as] one of the greatest slave-trading centres in Africa."[55]

Yet, for all this trading activity, coeval references to firearms in the Maseko heartland remain very sparse.[56] Two days south of Chikusi's capital, Kerr remarked that the Ngoni "had not a good show of flint-locks." Indeed, the "large" impis whom he met as they were making their way back from the Shire Highlands, the Maseko's principal raiding ground, were only "armed with bows and arrows, assegais and shields."[57] Elsewhere, the same Kerr noted that the knobkerrie was the Maseko Ngoni's "universal weapon. It was carried by every male; and was, no doubt, the implement most used in slave raiding, in conjunction with the shield." Shields themselves were a "terrifying emblem of war" that struck "horror to the hearts" of Ngoni victims.[58] In Chikusi's gun-poor capital, heavily armed foreigners stood a better chance of impressing their hosts than did less well-equipped visitors. While the lonely Kerr was treated with diffidence, the Portuguese Eustaquio da Costa, who moved about with a large number of armed Chikunda elephant hunters, enjoyed Chikusi's "respect, if not affection."[59]

The Yao trading principalities along the southern and southeastern shores of Lake Malawi were the other key commercial partners of the Maseko in the mid-1880s. In 1885, caravans from the Yao of the *Makanjira* were seemingly wont to visit Chikusi's.[60] In 1886, previous tensions between the Maseko leader and the Yao of the *Mponda*, installed in conquered Mang'anja territory since the 1850s, were momentarily resolved,[61] and the newly acceded Mponda II, Nkwate, became a conduit through which "many of the prisoners captured during the raids of the Angoni" found their way to the east coast.[62] In virtue of his sustained "intercourse with Swahilis and coastmen," Mponda II had access to a virtually "unlimited supply" of powder and arms,[63] including probably "a goodly number of Enfield rifles."[64] Yet, noted British Consul Hawes, who visited Chikusi's capital in June 1886, only "a few inferior muzzle-loading fire-arms" had found their way to the Dedza Highlands, and these were not being "used in warfare," "spears and clubs" being "the arms in common use" among the Maseko Ngoni.[65] King Chikusi, confirmed Last, Hawes's traveling companion, did have "a few guns, but it appears they are never used in the raids upon the neighbouring tribes, but for elephant hunting, or occasionally when parties are sent on duty to a neighbouring territory, in which case a gun or two is taken, probably for the purpose of firing a friendly salute on arrival."[66]

In 1876, as mentioned above, Young, the leader of the Livingstonia expeditionary party, met a Ngoni impi at Matiti, near the Shire's cataracts. The naval officer was clearly impressed, and he thought that the Zulu-like "discipline" of the Maseko Ngoni and tendency "to fight at close quarters with the assegai or spear" gave them an advantage over people armed with bows and arrows or "the miserable trade musket of the coast."[67] In reality, once more, there is enough evidence to suggest that the Maseko military superiority was not fully entrenched in southern Malawi, both at the time of Young's visit and in the following decade. The well-armed "Kololo" chieftainships of the Shire Valley were seemingly regarded with respect by the Maseko Ngoni and only rarely attacked.[68] Indeed, wrote the missionary-turned-planter John Buchanan, it was largely because of their gun-driven ability to provide a degree of protection to harassed Mang'anja that Livingstone's former associates succeeded in imposing their sway over parts of the valley.[69]

Especially telling are the dynamics of the Maseko's involvement in the civil war precipitated by the accession of the aforementioned Nkwate Mponda II in 1886, by which time the division of the Maseko into two separate and hostile sections—one under Chikusi and the other under his cousin, Chifisi—had reached the point of no return. Nkwate's succession to the position of *Mponda* was challenged by his half-brothers Chungwarungwaru and Malunda. From 1888, the two Yao factions were locked in a state of low-intensity but incessant war.[70] The Yao and the Ngoni succession disputes soon coalesced, for while Nkwate's internal enemies drew on the assistance of Chikusi's warriors, the new Mponda himself enlisted the support of Chifisi's regiments.[71] Both Chifisi and Chikusi's mercenaries were lightly armed.[72] It is thus not surprising that during serious confrontations the rich arsenal of Mponda II—in whose heavily fortified village even "ten-year-old children proudly [bore] rifles"![73]—always got the better of Ngoni impis attacking in formation.

The Roman Catholic White Fathers who spent almost two years at Mponda II's have left a graphic description of one such encounter. On 22 January 1891, the day of a joint Yao-Maseko attack, Mponda II, having made the required offerings to the spirits, sat "radiantly" inside his stockade. "At his side were seven barrels of powder and a stack of musket balls piled up on a mat."[74] These were meant to provision a force of about one thousand defenders.[75] The Catholic missionaries observed the advance of the Ngoni, unmistakable because of their "massive headdresses," with their field glasses. "The enemy are moving in, slowly and inexorably, destroying the fields of sorghum as they go. Finally a fusillade from the southwest corner of the town. It is 10:30 a.m. Five minutes later the 'lou-lous' of the women; the enemy has fled at the first volley, carrying off their wounded and leaving four dead behind them. Four bloody heads are soon brought into the village. What cowardice for the redoubtable Mangoni!"[76] By March 1891, with the Maseko unpreparedness vis-à-vis firearms having been so starkly exposed, Mponda II's people were "no longer even mildly frightened of the terrible Mangoni," and had indeed begun to raid the Dedza Highlands.[77] Late in April, the capital of Chikusi—who would die shortly thereafter—was ransacked by Mponda II's gunmen.[78]

Mpezeni's Ngoni

After settling athwart the contemporary boundary between eastern Zambia and western Mozambique c. 1860, the Ngoni of Mpezeni I, too, had several chances to experience the power of firearms. In 1863, what is very likely to have been one of their impis failed to storm the stockade of Chinsamba's, a Mang'anja settlement on the Linthipe River, to the east of present-day Lilongwe. Livingstone, who visited the village on the day after the attack, discovered that "Chinsamba had many Abisa or Babisa in his stockade, and it was chiefly by the help of their muskets that he had repulsed the Mazitu."[79] During the same decade, Mpezeni also came up against the guns that Chikunda "transfrontiersmen" (to use the Isaacmans' term) were occasionally putting at the disposal of harassed Chewa and Nsenga communities.[80] Indeed, the alliance between the Nsenga of the *Mburuma* and the Chikunda of Luso-African adventurer and state-builder Chikwasha may have been one of the factors behind Mpezeni's decision to shift the heartland of his kingdom to Chipata between the late 1860s and the early 1870s.[81]

Adopting M'Mbelwa's own strategy for dealing with heavily armed neighbors, Mpezeni himself enlisted the services of Chikunda elephant hunters against the ruling Mkanda, a Chewa chief who may initially have invited the Ngoni into his own territory, but who would eventually be overrun by the new arrivals—and their hired guns—in about 1880.[82] The deceased Mkanda's heir and some of his subjects took refuge with the then Mwase Kasungu, whom the eastern Zambian Ngoni, just like their northern Malawian counterparts, would always fail to subdue on account of his impressive defenses and arsenal. In 1889–1890, Mwase Kasungu's troops could apparently muster "more than three thousand firearms."[83] Firearms—as is made clear by the German-born ivory trader and diarist Carl Wiese—were by this stage a central feature of Mwase Kasungu's capital. While "almost all" of its numerous inhabitants had access to them, local metalworkers—including Chibisa, the chief's brother—found the technology of even Wiese's "modern weapons" intelligible and accessible.[84] This being the case, it is scarcely surprising that Kasamba Malopa, the important *induna* of Mtenguleni, one of the two capitals of Ungoni, should have taken a leaf out of Ng'onomo's book (see above) and shown little willingness to confront so well-armed an enemy.[85]

To be sure, Mwase Kasungu's position was unique. No other Chewa chief could equal his resources and military preparedness vis-à-vis the Ngoni. Ngoni raids, indeed, forced the southern Chewa and their leaders, the most notable of whom was the then holder of the ancient *Undi* title, to seek refuge on mountainous retreats, where they eked out "a miserable existence . . . suffering almost every year from famine."[86] Yet, even in Maravia, Édouard Foà noted with reference to the year 1891 that Mpezeni's people were still "frequently defeated," as they were only armed with "bows, shields, assegais and knobkerries," whereas the then Undi's people had guns and powder.[87] At this point, it is important to reiterate that muzzle-loaders were not always superior to Ngoni shields and such *armes blanches* as assegais and knobkerries. They became so, however, when associated with such strongly fortified positions as both Mwase and Undi evidently made use of in the last decades of the nineteenth century. This is aptly borne out by the fate that befell a tactically naive party of "Arab" traders on the upper Bua River in 1887. Having been lured out into the open by a youthful Ngoni advance regiment, the traders, who believed "these to be their only enemies," fired "their weapons almost all at the same time" and began to give chase to the retreating impi. They soon fell into the hands of two more regiments, who lay in ambush. "The Arabs were then attacked . . . from the rear. Not having time to reload, they began to be massacred by club and *assegai*."[88]

Victories such as this may have contributed to the Zambian Ngoni's choice not to deploy firearms for military purposes. The use of the word "choice" is appropriate, for, in this instance too, what we are confronted with is a deliberate decision, rather than the enforced consequence of the lack of economic opportunities. The aforementioned Wiese's first visit to Mpezeni dated to 1885–1886. The establishment of a semipermanent trading base in Mtenguleni followed thereafter.[89] By then, other merchants were active in the area. These included both Zambezian entrepreneurs—such as the Portuguese Joaquim Augusto do Rego, whose presence in Ungoni was reported in 1889, and the Luso-African Francisco Jose Pacheco, active among the Nsenga and neighboring people since 1884[90]—and Arab-Swahili traders, who, according to Wiese's probably exaggerated description, "continually visit[ed]" Mpezeni, in spite of the dangers posed by some of the same king's less disciplined regiments.[91] During Wiese's stay in Ungoni in 1889–1890, at least one Zanzibari caravan made Dingeni its temporary headquarters while waiting to "buy slaves and ivory."[92] While there is little doubt that Mpezeni sought to monopolize the trade in ivory, of which he sometimes "had a great deal,"[93] the evidence relating to his involvement in the slave trade is more contradictory. The Arabs of Dingeni did undoubtedly acquire some slaves.[94] Yet they were also reported to have "sold slave women to Mpezeni," who, Wiese remarked, "[bought] people but never [sold] them."[95]

The primary Ngoni import was cloth, which by the late 1880s was becoming common both among women, "especially the rich ones," and men, some of whom were "beginning to dress as Portuguese subjects."[96] King Mpezeni himself had "many suits and clothes of good quality."[97] But firearms also featured prominently among Mpezeni's possessions. Wiese, in fact, described him as an enthusiastic "collector" of weapons "of modern workmanship." These guns, however, were deployed solely as symbols of royal wealth, for they were "never use[d]."[98] What was true of Mpezeni was also apparently true of his people as a whole. "The Ngoni," Wiese declared, "have remained very conservative regarding their armament. Although nowadays many Ngoni already possess firearms, some two or three thousand in my computation, they do not use them in war to the fullest extent. They use, as in the early times, wooden clubs, *assegais* (they carry two or three of them), and shields, which are elliptically shaped and made of oxhide."[99] Hunting, too, was carried out without the aid of firearms, as the Ngoni were "not experienced enough" in their use.[100] The military propensities of the Jere Ngoni of Chipata remained unchanged over the course of the next few years, for in April 1896 Lt. Col. R. G. Warton, of the North Charterland Exploration Company, could still write that Mpezeni's forces were "unaccustomed to handle" even the "obsolete guns" to which they had access, "preferring assegais and bows and arrows."[101] Warton's first meeting with Mpezeni confirmed the validity of the information he had initially received. The Ngoni—he wrote in August of the same year—"do not rely very much on firearms in case of war, almost invariably using their assegais and knob-kerries."[102]

EXPLAINING TECHNOLOGICAL DISENGAGEMENT

The few regional specialists who have pondered over the Jere and the Maseko Ngoni's reluctance to adopt firearms for military purposes have tended to accept Margaret Read's old statement to the effect that the "antique guns" put in circulation "by the Portuguese and Arabs were no match for the Ngoni skill with the long throwing-spear or the short stabbing-assegai, and the bullets were such that the stout cow-hide shields of the Ngoni could withstand them."[103] In light of the numerous military setbacks suffered by every major Ngoni group in the second half of the nineteenth century, this minimalist, deterministic explanation is either wide of the mark or, at best, able only to tell a very limited part of the story. I contend that a closer investigation of Ngoni sociocultural institutions is needed if we are to make sense of the military conservatism described in the previous section.

Ngoni armies were made up of age-set regiments open to all able-bodied youths and men. While the ultimate origins of this principle of social and military organization are to be found in developments that had affected KwaZulu-Natal, the homeland of the leaders of the initial wave of migrations, in the late eighteenth century (see chapter 1), the meritocratic, achievement-oriented aspects already inherent in the Nguni archetype were enhanced by the experience of migration, during which the need forcefully asserted itself to incorporate—and secure the allegiance of—large numbers of foreign captives. Thus, successful participation in the Ngoni military machine was made to serve the twin purposes of bringing about a common (though by no means fixed or irreversible) identity and favoring social promotion for scores of captives, who soon formed the majority of the members of all of the Ngoni polities in the central African interior.[104] In the 1930s, the northern Ngoni historian Chibambo described the process in the following terms: "In [war], ordinary people and slaves came together. . . . The slaves who showed their courage and strength in war quickly received their freedom and many also had villages of their own because of the people they had captured in war; others obtained the standing of men in authority."[105]

To be sure, Chibambo's romanticized account papers over the many internal tensions which emerged after the northern Ngoni settled in Mzimba and which expressed themselves in, for instance, the Lakeside Tonga insurgency of the late 1870s. Yet there is no need to reject its essential contours, which are confirmed by several other sources. The missionary Fraser, in Mzimba since the mid-1890s, found out that, besides "civil indunas," who "were men of great power, and were all emigrants from south of the Zambesi," the late M'Mbelwa I's kingdom had also comprised many royally appointed "war indunas, some of whom were originally slaves, and belonged to the local Nyasa tribes."[106] The situation obtaining among the Ngoni of Mpezeni was much the same, for Wiese wrote that prisoners of war were not only used "on agricultural work and other services," but were also "adopted later into Ngoni customs, language, and dress and sometimes reach[ed] important positions." He knew "many who [had] influence and wealth."[107]

Among the Ngoni of both Zambia and Malawi, promotion within the regimental structure went hand-in-hand with the possibility of partaking of the patronage networks through which raided captives and cattle were distributed among deserving members of specific army units.[108] In the late nineteenth century, Ng'onomo, whom we have already encountered in the previous section, was the living embodiment of the opportunities for individual achievement thrown open by a system in which military skills were valued as much as—if not more than—birth. Originally a Thonga captive from Delagoa Bay, Mozambique, Ng'onomo, having "proved himself a warrior braver and more successful than most," had been remunerated by M'Mbelwa "with wives, slaves and cattle," and he had eventually risen to the position of overall leader of the northern Ngoni army.[109] It was perhaps with a view to regulating their rewards mechanisms that the Ngoni of both M'Mbelwa I and Mpezeni I developed formal means to honor military bravery. A kind of "decoration, or a military order," consisting of "enormous ox horns," was in use among both groups,[110] while M'Mbelwa also bestowed the title of "Master of the Stockade" on the first fighter to have scaled the palisade of a besieged village and slain one of its inhabitants. When brought before the presence of M'Mbelwa, the hero of the day carried "in his hand the bow or gun of the man he [had] killed," danced, and was given a bullock "as a signal token of [the king's] princely admiration."[111] Soldiers returning from war were split into two groups, for only those who had "blooded their spears" were permitted to parade before the king and be individually praised and rewarded for having fought with distinction.[112]

Fortitude and the determination never to yield to the enemy were central themes of the *imigubo* war songs chanted by the regiments on the eve of a fight. "Let us strike, they have given up" was the refrain of one of the imigubo collected by Read at Emcisweni, Mperembe's old capital, in the mid-1930s.[113] "I have been longing to sleep / The tremendous size of their villages / The wounds of the spears"—went part of another song from Mzimba.[114] Cowardice and the tendency to bring battles to a premature end, conversely, were openly berated in a war chant common to both the Jere and the Maseko Ngoni of Malawi:

> The cattle run away, you cowards.
> Those yonder; they run, you cowards.
> The cattle, see, do they run? They run, you cowards.
> Is your young manhood over? They run.
> You are left with the carriers. They run.
> Look at the cattle, they run, they run;
> You have eyes only for the food-stuffs. They run.[115]

Other imigubo advocated stoical acceptance of death, as did a long song from northern Malawi, only the first stanza of which I reproduce below.

> The earth does not get fat. It makes an end of those who
>   wear the head-plumes.
> We shall die on the earth.
> The earth does not get fat. It makes an end of those who
>   die as heroes.
> Shall we die on the earth?
> Listen, O earth! We shall mourn because of you.
> Listen, O earth! Shall we die on the earth?[116]

Military bravery, then, was the key to unlock the potential for self-advancement inherent in Ngoni institutions.[117] But notions of "military bravery"—an important component of John Iliffe's category of "heroic honour"—are far from universal and always depend on a given social group's approved codes of behavior.[118] Insofar as the Ngoni military is concerned, the practice of hand-to-hand fighting with spear, knobkerrie, and shield was precisely one such prescriptive, hegemonic norm. The "right to command respect" among the Jere and Maseko Ngoni, and to reap the social and material advantages of such respect, was forged in the heat of close combat, which—the Maseko Ngoni told Last in 1886—did not involve the use of missile weapons. Rather, "on coming to close quarters, [Ngoni fighters] strike their opponents' legs, and when they have brought them down, then spear them."[119] It is not for nothing that one of the local names for the Ngoni in the territory corresponding to present-day northeastern Zambia was "the stabbers" (*bachime*).[120] Unlike in nineteenth-century Zululand or Buganda, no Ngoni voice is on record as having described the gun as the "coward's weapon."[121] Yet there is no doubt that not all forms of warfare were construed by the Ngoni as equally commendable and respectable. It was because they undermined a preexisting warrior ethos and tainted dominant notions of honor that firearms, their manifest military usefulness notwithstanding, were regarded with "disdain" (dédain) by the Ngoni of both Mpezeni and Chikusi.[122] Hence, the most that could be done with them was to leave them in the hands of independent foreign allies and mercenaries (such as the aforementioned Chikunda, Senga, and Mwase Kasungu's Chewa), who had no stake in the infra-Ngoni competition for upward mobility and who could therefore afford to ignore the demanding requirements of Ngoni heroic behavior. The Zulu, after all, had opted for a similar course of action at the time of Shaka and Dingane.[123] It was only during the Anglo-Zulu War of 1879 that Zulu tactics changed so as to incorporate firearms to a greater extent than most specialists have been prepared to concede. Even then, however, the gun remained culturally subordinate to the assegai, the epitome of "Zulu ideas of the Zulu way of war."[124]

As in the case of the Zulu, firearms never became central markers of masculinity for the Ngoni. Rather, it was the stabbing spear and the knobkerrie that most readily epitomized the bellicose qualities of the ideal Ngoni fighter. Imigubo celebrated the spear's ancient origins, and the wealth and the military triumphs it had made possible:

No Paramount can be poor because of the spear
Then why are you running away?[125]

～

Fight now! Come and fight now!
Slay them! We'll brandish spears!
Straight forth doth speed your arrow.
Tremble! Yes! They tremble!
    When we draw near,
      And far they'll flee as we approach them!
Sharpen keen your arrows!
Brave heads upraised and shouting
    Loudly our defiance.
All there who oppose us
    Quickly our spears
Shall pierce their breasts. They will be scattered.[126]

And it was with spears that lions, the most dangerous and prestigious of animals, were killed—as attested by Deare, who in 1896 or 1897 was specifically asked by some of Mpezeni's soldiers not to shoot a cornered lion, as they wished to show him how they fought the beast. "Forming a ring the natives started in to worry him; first one and then another advancing tendering him the shield to jump at. If he did so the shield was dropped and a quick lunge was made with the assegai, the man cleared, and another did the same thing, until the lion exhausted from loss of blood, collapsed. . . . It was quite a unique performance, reminding me of a bull fight."[127]

The celebrations that followed the killing of a cattle-eating lion at M'Mbelwa's in 1879 provide some insights into the masculine warrior ethos that prevailed among the northern Ngoni—and into the central position that edged weapons occupied in it. After laying down the lion skin, Laws reported,

> the warriors marched along and formed three deep, all resting
> their shields on the ground, and presenting rather a formidable
> appearance, nothing being seen in front except a wall of bullhide
> and rows of heads above it. In turn each of those who had had a
> share in killing the lion sprang forward, giving a shout, to which
> his companions at once supplied a chorus, and kept up a sort of
> dance, stamping their feet, and shaking their spears or beating

their shields with their clubs. Meantime a number of the young women of the tribe, gaudily dressed with beads and coloured cloths, kept moving backwards and forwards in front of the warriors, carrying a slender rod some 10 ft long in their right hand, to which they imparted a trembling motion, and at the same time jerked their heads about as if they meant to pitch them off. After this had gone on for some time, the poet of the tribe stood forth, with spear and shield, and recited in the hearing of the chief and other headmen how the lion had been killed.[128]

Among the Ngoni of Malawi, the spear was the only personal object not to be buried alongside the body of a deceased individual, being used instead as a relic and "to facilitate communications with the ancestors."[129]

The struggle for status and self-improvement—one that expressed itself in a high degree of personal investment in the core values of Ngoni militarism—was principally the affair of the youths, from whose ranks the majority of captives were taken. Both coeval observers and later scholars have stressed the significance of intergenerational cleavages among various Ngoni groups and the pressures exerted on Ngoni leaders by their younger followers. In particular, while senior politico-military chiefs, having already attained a high position in society, were not averse to regulating—if not abolishing altogether—warfare and raiding with a view not to antagonize the encroaching Europeans (be they missionaries, as in Mzimba, or concession owners and administrators, as in Chipata), younger warriors belonging to newly instituted regiments resisted this tendency on account of their eagerness "to obtain wealth, or wives and slaves."[130] Unlike their seniors, junior soldiers "felt that unless they established their social credentials in war—'washed their spears in blood'—the system into which they were trying to gain a substantial hold would have no meaning or significance."[131] The result, as will be seen in the next chapter, was that a good many of the raids witnessed by early European observers on the outskirts of the major Ngoni kingdoms of both Zambia and Malawi represented local—as opposed to centrally organized and sanctioned—initiatives, and that it was most probably the youths, rather than their elders, who acted as a check on military innovation, insisting that these same wars be fought with obsolete, but prestige-enhancing, edged weapons and shields.

The openness of the Ngoni social systems and their relationships with specific notions of heroic behavior and masculinity illuminate the

FIGURE 5.1 Members of a youthful Ngoni regiment. Source: Édouard Foà, *Résultats scientifiques des voyages en Afrique* (Paris: Imprimerie Nationale, 1908).

root cause of resistance to firearms, particularly (and paradoxically) among such groups as, in more rigidly stratified societies, are commonly viewed as the natural supporters of "progress" and "change"—including technological change. Since their ability to rise to positions of greater power and influence depended on the extent to which they succeeded in proclaiming their "right to respect" by excelling in warfare and socially prescribed fighting methods, young warriors of captive origins cannot have been keen to drop their hard-won military skills with a view to engaging with an exogenous technology—firearms— which did not offer comparable prospects of material rewards and which might indeed have hampered their progress towards full adulthood and social acceptance.

↬

The conclusion to which our discussion leads is that there is much to be said in favor of the old functionalist topos that upwardly mobile societies produce more conformists than revolutionaries. In the case of the Jere and Maseko Ngoni, of course, this trait manifested itself with particular clarity in the military sphere, where opportunities for self-improvement were most readily available. The reaction to externally introduced technology—be it positive, as in the examples examined in the second part of this book, or negative, as in the present case—is always contingent and informed by preexisting social circumstances. In the case of the Ngoni, such circumstances militated against the widespread adoption of firearms for war purposes. Lozi royals, Luvale and Kaonde hunters, and Yeke market-oriented conquerors all took up the new technology because—in a suitably refashioned form—it could be made to serve a distinctive set of contextual interests. Ngoni fighters of captive origins rejected it because its appropriation would have subverted the very foundations of the social systems which they were intent on scaling and which appeared to offer them enough opportunities in their present form. This technological disengagement was not without consequences. It is to these latter that we first turn in the next chapter.

6 ᦒ Of "Martial Races" and Guns
The Politics of Honor to the Early Twentieth Century

OF THE three Ngoni groups examined in the previous chapter, only Mbelwa's avoided a violent clash with European-led armies at the outset of colonial rule. This was partly a consequence of the intermediary role played by the Livingstonia missionaries, the only European residents in the northern Ngoni's heartland between the late 1870s and the beginning of the twentieth century, and partly a result of the lack of significant European economic interests in the area.[1] Conversely, the subjugation of the Maseko Ngoni and Mpezeni's Ngoni was a bloody affair. The first section of this chapter focuses on the causes and nature of these military confrontations. Its main purpose is to foreground the endurance and ultimately devastating effects of what a colonial historian once termed the Ngoni "cult of cold steel."[2]

The second section of the chapter explores the impact of conquest on Ngoni militarism. It will be argued that, far from spelling the immediate end of Ngoni notions of military honor, colonialism provided an opportunity for the recasting and, to a degree, survival of such notions. In this respect, my approach heeds Thomas Spear's call not to overstate colonial ability to invent and manipulate African identities and institutions.[3] The imperial frontier—this chapter and the scholarship that inspires it argue—was never a clean slate onto which European administrative and military leaders could inscribe whatever meanings suited their hegemonic project. Like European technological artifacts, European representations and discourses—including the so-called "martial race ideology"—were never left unaffected by the encounter with preexisting indigenous societies, structures of meaning, and material forces. However, insofar as the central subject of this book is concerned, the effects of European conquest *were* revolutionary. Construed as a "martial race" partly on account of their reliance on edged weapons and devotion to the ideal of close combat, the Ngoni were recruited in substantial numbers into colonial police forces, one of the distinguishing traits of which was, paradoxically, the right to bear firearms. Under the new circumstances, then, the gun became everything it had not been in the precolonial context, replacing the assegai as a central symbol of Ngoni masculinity and major route towards individual improvement.

THE SUBJUGATION OF THE NGONI
Mpezeni's Ngoni

The history of Ngoni-European relations during the decade or so that preceded the war of January 1898 has been told in considerable detail before.[4] For our purposes, a chronological summary suffices that highlights the motives of the protagonists and the structural causes of the conflict. Between the summer of 1889 and the spring of 1891, a Portuguese expedition resided in Mtenguleni, one of Mpezeni I's two capitals in Ungoni. Its unofficial leader, Carl Wiese, a German-born ivory trader and *prazeiro* along the Zambezi, became very close to the Ngoni king, whom he had already visited twice in the recent past. Indeed, on 14 April 1891, a few days before the departure of the last members of the expedition, Mpezeni was persuaded to sign a concession that granted Weise — rather than Portugal — extensive mining and other rights in the heartland of the kingdom.[5]

For the next few years, notwithstanding the Wiese concession and the widespread belief that Ngoniland was rich in gold, Mpezeni's Ngoni were left undisturbed by Europeans. The Anglo-Portuguese treaty of June 1891 had placed the bulk of Mpezeni's country in British territory. But in the early 1890s, Cecil Rhodes's chartered company, the British South Africa Company (BSAC) — to which the administration of the British sphere to the north of the Zambezi had been delegated — was in no position to even attempt to penetrate Ungoni, and it limited itself to funding the establishment of a few small stations on the outskirts of Bemba and eastern Lunda territory. Meanwhile, the infantry forces of the British Central Africa Protectorate (BCAP) (Nyasaland), ratified in 1891 under pressure from Scottish missionary interests, were being kept fully occupied by prolonged Yao resistance in the southern part of the country.

The first European representatives seriously to affect Ngoni politics were the prospectors of the short-lived Rhodesia Concessions Company (RCC). Fort Partridge, the strongly stockaded post that they occupied between the end of 1895 and mid-1896, was located in the territory of Mafuta, a Chewa tributary of the Ngoni, some twenty-file miles to the north of Luangeni, the village of Mpezeni's senior wife.[6] The military resources available to the RCC were considerable and seem to have included two Maxim guns, one of which was at one point reportedly used to "impress" the "natives" in the surroundings of Fort Partridge.[7] Such displays of force must have disturbed the Ngoni leadership, who probably suspected that Mafuta and other exploited Chewa might seek to ally themselves with the Europeans with a view to shaking off their subordinate status.[8]

At exactly the same time that the Ngoni hierarchy was mulling over the ultimate intentions of the RCC, the armed forces of the Protectorate stormed Mwase Kasungu's fortified town in the context of their anti-Yao operations.[9] As has been seen in the previous chapter, Mwase Kasungu, who committed suicide following his defeat, was one of the few Chewa leaders whom Mpezeni's Ngoni had always failed to subjugate. Understandably, the Ngoni king regarded Mwase Kasungu's demise as a threatening development. A further unmistakable sign that the tide of European occupation was fast approaching Ungoni was the establishment, in mid-1896, of the first BSAC station in the future Eastern Province of North-Eastern Rhodesia. Fort Jameson—as the post was known until 1898—was sited among the Chewa of the *Chinunda*, some fifty miles to the north of Luangeni. Though Warringham, the founder of the station, and his meager force of twenty-five police had been urged to obtain the "good will and friendship of Mpezeni,"[10] the very existence of the Maxim-equipped *boma* and the fact that it fell well within the Jere Ngoni's traditional raiding territory are likely to have compounded their anxieties about future intercourse with the encroaching *vishanzi* ("people from the sea"[11]).

In mid-1896, the RCC was superseded as the representative of European business interests in eastern Zambia by the North Charterland Exploration Company (NCEC), which had obtained a large land grant and related mineral rights from the BSAC on the strength of Wiese's 1891 treaty with Mpezeni.[12] Relations between the Ngoni hierarchy and the NCEC's party, initially led by Lt. Col. R. G. Warton and including a sizeable private police force raised by Wiese in Nyasaland, did not start on a hostile footing. Employing Wiese as go-between and deemphasizing the British nature of its expedition, the NCEC was granted permission by Mpezeni to build a base (known as Fort Young in 1897–1898) in Luangeni and to carry out a series of prospecting expeditions in surrounding areas.[13] Mpezeni clearly hoped to use the NCEC as a bulwark against the feared administration of the BCAP, which had already given abundant proofs of its powers, and the BSAC, which the Ngoni king probably regarded as a mere extension of the Protectorate.

Yet tensions were not long in coming to the surface between a more and more easily identifiable war party, led by Mpezeni's son and heir, Nsingo, the army commander whose compound was located in the immediate proximity of Fort Young, and NCEC employees, under acting manager Major G. R. Deare, between September 1896 and mid-1897. As early as October 1896, Forbes, the Blantyre-based BSAC administrator, had reported that NCEC prospectors were "living . . . at the pleasure" of Mpezeni, since "a large portion of the people [were] very averse to whites settling in their country."[14] Deare, too, became increasingly aware of what he euphemistically called "signs . . . of disaffection" towards him, his fellow Europeans, and their African employees.[15] Such hostility was closely related to Nsingo's growing awareness that the presence of Europeans and their armed retainers in Ungoni and surrounding areas would eventually spell the end of such raiding systems as had hitherto enabled the fighters of the age groups he represented to prove their military prowess in battle and, in so doing, enhance their standing in Ngoni "meritocracy."[16] Having already attained a high socioeconomic status on the strength of their past military performance and display of heroic honor, politico-military leaders of Mpezeni's generation could at least contemplate the possibility of compromising with the white intruders and yielding to the latter's opposition to continuing warfare. Conversely, from the point of view of younger soldiers, the NCEC and the other Europeans posed an unacceptable threat to a way of life and system of social relationships the benefits of which they had not yet fully reaped.[17]

From the beginning of 1897, Mpezeni struggled to keep the lid on pressures emanating from the war party. Still, unsanctioned raids on the periphery of Ungoni were reported on several occasions.[18] The traveler Hugo Genthe witnessed the immediate aftermath of one such foray near Fort Jameson in May 1897. "On my arrival [at Chuaula's] I found the male population all under arms, and the women crying. A raiding party of Mpeseni's people had attacked them suddenly that morning. Ten women were killed in the gardens and twenty-two were taken away as prisoners. An old man and one of Chuaula's children had been very severely wounded. Their entrails hung out frightfully torn wounds, inflicted most likely by barbed spears."[19] Genthe was a perceptive observer, and he grasped the social logic behind the militancy of assegai-wielding Ngoni or Ngonized youth: "To understand their motives one must know that the young Angonis are not allowed to participate at the big games unless they already have 'wetted their spears,' that is, killed somebody."[20] The same traveler provided the following graphic illustration of the significance of intergenerational cleavages in Ungoni and of the manner in which such tensions expressed themselves in the adoption of contrasting attitudes towards the vishanzi. En route between Luangeni and Chimpinga, where Mpezeni was then residing, Genthe, who was being carried on a hammock (machila), stumbled across

> a party of ten Angonis—some headmen accompanied by a few armed men. The path was narrow. . . . I expected the natives would step aside to let pass the machila, but nothing of the kind happened. When my men requested the Angoni politely to stand aside, they pointed out silently with their knobkerries into the mapira [sorghum] field, and my machila men took the hint. . . . I swallowed the afront [sic]. A little further on we met another party—one of Mpeseni's indunas and his men. This old fellow stood aside to make room for the machila and saluted me. As his behaviour was a contrast to that of the headman whom I had just passed, I questioned him about the latter's name, and found that it was Mpeseni's eldest son: I did not wonder any more at his insolence.[21]

Wiese, who knew the Ngoni better than any other European and understood the difficulties faced by Mpezeni and his elders in restraining junior regiments, thought the BSAC would do well to raise a force of 200 to 250 police and dispatch it to Ungoni with a view to strengthening the position of the Ngoni king and assisting him in enforcing the ban on raiding he was said to have decreed.[22]

As noted by Langworthy, had Wiese's suggestion been taken up, it might well have altered the future course of events in eastern

Zambia.[23] It was not. On the contrary, the defeat and execution of Gomani, Chikusi's successor, about which more will be said below, coupled with Wiese's absence from Ungoni throughout most of 1897, helped sway opinion in favor of war. The assertiveness of Nsingo's followers, evidenced by increasing levels of violence on the outskirts of the kingdom, led the BSAC, which could now count on the promised support of the Central African Rifles (CAR)[24]—the name of the Protectorate's armed forces since the mid-1890s—to change its hitherto prudent approach. By the end of 1897—when "hardly a month [went] by without several raids being reported"[25]—Warringham, the Company's "collector" in Fort Jameson, and his superior in Blantyre, Acting Administrator Daly, had reached the conclusion that the Ngoni had to be "smashed" and "dealt with" once and for all, "if in the future the BSA Co. wishes to collect labour in these districts for any purpose whatsoever."[26] However, the playing out of intergenerational tensions in Ungoni continued to the end, for it is said that, late in 1897, when Wiese and three other European employees of the NCEC were besieged in Fort Young by Nsingo's spearmen—a development that prompted the BCAP administration to dispatch a large-scale military expedition across the border—Mpezeni still sent provisions to the beleaguered party.[27] Mpezeni and the old guard probably still hoped that a rapprochement might be effected, and this uncertainty as to the final outcome of the power struggle in Ungoni may account for the fact that, when war with the vishanzi finally came about, the Ngoni were actually unprepared for it.

With no national army mobilization having been ordered, and with the bulk of Ungoni's population distracted by the yearly *nkhwala* celebrations,[28] the Ngoni were slow to react to foreign aggression. The first CAR contingents reached Fort Jameson during the first half of January 1898.[29] Conflicting reports about the fate of Wiese, still hemmed in at Fort Young, convinced their commanding officers to advance towards the Ngoni heartland without awaiting the arrival of Lt. Col. W. H. Manning—the CAR's commandant and acting British commissioner in the Protectorate—and his additional troops.[30] The invading force consisted of about 350 African and Sikh regular soldiers and 200 porters under the general command of Captain Brake, who was assisted by three fellow European officers and Warringham of the BSAC. Accompanying the troops were two Maxim machine guns, two 7-pounder RML mountain guns, and over twenty thousand rounds of rifle ammunition.[31] Brake's column left Fort Jameson on 18 January 1898, reaching the abandoned Fort Partridge on the evening of the same day. No contact with the Ngoni was made during the march, despite the progression of the British force being hampered by heavy rain, and despite it having to pass through several potential ambush positions that the Ngoni could well have taken advantage of.[32]

Once the CAR entered central Ungoni on 19 January, the default position of such regiments as could be mustered at short notice was to attack in the open, seeking to make direct contact with the enemy. The first significant encounter of the war took place to the north of Luangeni/Fort Young, when a two-hundred-strong impi schooled in the principles of honor-enhancing close combat was "seen advancing towards our left front. . . . As the enemy closed down, the left wing of the advanced guard fronted left, and, when the enemy were within thirty yards, poured in a steady volley which checked them, and, on the appearance of the flanking party in their rear, they drew off slowly, with a loss of some half-a-dozen men."[33] A later account explained that the Ngoni casualties consisted of the "bravest" of their numbers, who had "endeavoured to charge our line, but were shot before they got near."[34] A few hours later, "one mile from Loangweni," "some 150 Angoni from the village of Mpeseni's son Singu rushed down on the right flank of the column." Brake's A Company "halted and . . . delivered two volleys, which effectually checked the rush, and the Angoni retired."[35]

The key set-piece battles of the war took place around Fort Young in the morning and afternoon of 20 January. On both occasions, the Ngoni made very limited use of firearms. In the morning, the British faced a party of "600 or 700" warriors under the leadership of Mlonyeni, another son of Mpezeni. Mlonyeni's forces were apparently bent on retrieving the thousand head of cattle that the British had captured on the previous day. They advanced "dancing and brandishing their weapons within 300 yards." The CAR soldiers marched forward with fixed bayonets. "Fire was opened by the centre companies at about 100 yards distance, two volleys being poured in, and the advance resumed. This silent steady advance impressed the Angoni almost more than the volleys had, and, throwing their spears, they slowly gave way."[36] The CAR soldiers followed the retreating Ngoni, who were also shelled by the mountain guns, and burned a number of villages where resistance was encountered. "The total loss to the enemy, from prisoners' accounts, was about twenty killed and several wounded."[37]

The afternoon of 20 January witnessed what Rau has described as the "most deliberate and organized Ngoni attack" against the invaders.[38] Led by Nsingo, "three long lines" of warriors deployed in the Zulu-inspired horn-shaped formation strove to encircle the British force, which had returned to Fort Young after the morning battle. The Ngoni pincer movement, however, clashed on a wall of fire, for "the 7-pounder guns and Maxims were brought into action and the troops fired steady volleys at the advancing masses. The Angoni came on bravely, but as their losses increased, the lines halted and then began to retire slowly. . . . The order was at once given to charge and the Angoni broke and were pursued for some distance."[39] More villages—including Nsingo's—were razed to the ground. Brake himself was impressed by the "considerable courage" showed by the "Angoni Zulus" in "advancing to within 50 yards of our men." It was their technological conservatism, however, that was to blame for their defeat, for Nsingo's spearmen "could not face the volley firing. Guns amongst them are apparently rare and their fire most inaccurate and their spears never reached the firing line. In no case were they able to charge home."[40] "Every officer" involved in the expedition, in fact, recognized that if the Ngoni had "been armed with guns . . . operations would have been vastly more difficult." Without firearms, however, the Ngoni simply "expose[d] themselves in their endeavour to charge down on our troops."[41] Nsingo's routed army may have numbered between three and five thousand men, reservists included.[42] Conservative estimates put the number of Ngoni deaths at fifty, while no casualties were registered on the British side.[43] In reality, Ngoni losses are likely to have been higher, since "many dead and wounded [were] carried off" by their retreating comrades.[44]

During the following two weeks, the invading troops—numbering almost one thousand by 28 January and representing the "largest and most efficient" force ever collected by Europeans in the region[45]—were split into four mopping-up columns led by white officers. "Several smart skirmishes" were fought "in which some of the Angoni showed that the old Zulu spirit still lingers amongst them, as they charged fearlessly up to small parties of our men" armed with rifles.[46] Punitive expeditions and search and destroy operations—in the course of which all of the largest villages and fields in Ungoni were burned down, dozens of additional Ngoni spearmen killed, and thousands of cattle requisitioned—lasted until the capture and execution of Nsingo, episodes around which a significant corpus of historical tales would eventually flourish.[47] Nsingo died near the ruins of Chimpinga on 4 February. Mpezeni surrendered a few days later, as the fires that had destroyed his once "thickly populated," "splendid pastoral country" were still smoldering.[48] The old king, who, after all, had sought to avoid a direct clash with the Europeans, was spared his son's fate, being instead briefly interned in Mchinji, in Nyasaland, from April 1898.[49] By this time, the Ngoni were reported to be "thoroughly demoralized and convinced of the uselessness of attempting anything against troops armed with rifles and machine guns."[50] Rifles, machine guns, and artillery fire had brought to an end the independent history of Mpezeni's Ngoni. However, as will be shown below, the story of Ngoni militarism was not yet over.

Maseko Ngoni

The violent incorporation of the Maseko Ngoni into colonial structures can be dealt with more briefly, not least because the immediate causes and nature of the conflict were not, in essence, dissimilar from those which obtained further west, in Mpezeni's territory. Even in Dedza and Ncheu, European military intervention was brought about by uncontrolled raiding—directed, in this case, against former tributaries of the Maseko Ngoni who had fallen under the influence of the Baptist missionaries of the Zambezi Industrial Mission in the Livelezi Valley, begun to pay taxes to the Protectorate administration, and been drawn into the orbit of both Yao and European planters in the Shire Highlands. One significant difference, however, was that among the Maseko Ngoni the salience of the clash between compromising elders and militant youth was apparently more muted than in eastern Zambia. No doubt, this had much to do with the fact that Gomani—who had recently succeeded his father Chikusi as paramount of one of two opposing Maseko segments (see chapter 5)—was little more than an adolescent and thus still in need of proving his "capabilities in warfare before their people" and beefing up his material and human resources through raiding.[51] Gomani, in the pompous words of a visiting colonial official, was a "modern Rehoboam," who rejected "the advice of his aged councillors" and was "being carried on to war and rapine by his 'young bloods.'"[52] This being the case, the chances of a peaceful transition to European rule in south-central Malawi were always slim.

Between 6 and 8 October 1896, "several large *impis*" led by Gomani himself descended on Ncheu. In a last-ditch attempt to regain control of local labor supplies and defend the viability of the regiment system, Gomani's fighters, who appear to have been armed with both spears and muskets, stormed and burned numerous mission villages, "killing the inhabitants and capturing large quantities of slaves."[53] Hundreds of refugees sought shelter in both Gowa and Dombole missions,[54] which the Ngoni left unmolested, notwithstanding Gomani's obvious awareness of the threat that the missions posed to the Maseko war-based economy and social system.

Having by and large overcome Yao resistance, the CAR was in a position to react speedily. An expeditionary force equipped with one or two 7-pounder guns and comprising "58 Sikhs" and "198 trained native troops" was put together and sent to the Kirk Range under the command of Captain Stewart.[55] Gomani's capital was reached in the early morning of 21 October. During the brief battle that followed, the Maseko Ngoni displayed a greater familiarity with firearms than the Ngoni of Mpezeni would do shortly thereafter. Yet it is clear that the tendency to attack in formation and the search for honor-enhancing close combat still dominated their military thinking. According to eyewitness R. C. F. Maugham, then still known as R. C. F. Greville, as soon as the invaders were spotted, Gomani's warriors, "well over a thousand strong," "formed a long line of considerable depth slightly advanced at the flanks, and proceeded to push forward across the interval of open space which separated us to repel our attack." Despite artillery fire creating havoc in their ranks, the Ngoni "continued steadily to advance, brandishing their weapons and clashing them together with a volume of sound which, added to their deep-chested war-song, was wild and stirring in the extreme."[56] For a time, Gomani's men "kept up . . . a hot fire" with the muzzle-loaders that "many" of them possessed, but the continuous rifle volleys of the CAR soldiers and the incendiary shells of their mountain guns eventually produced the desired effects, cutting short the Ngoni advance "at a distance of about 100 yards."[57] Scattered and demoralized, the Ngoni retreated towards their burning capital before accepting the enemy's bayonet charge.

Maugham's account of the battle suggests that, however numerous, firearms were still being used by the Ngoni of Gomani as a mere improvement on throwing spears, the hurling of which sometimes anticipated the final charge leading to hand-to-hand combat. As among the Zulu before the Anglo-Zulu War of 1879, the culturally subordinate status of firearms meant that the practical effects of their adoption remained limited. Firearms, in other words, did not usher in significant changes in military tactics or replace the assegai as the Maseko Ngoni's central fighting weapon. On 23 or 24 October, Gomani, who had fled the capital with a few hundred partisans, was tricked into attending a meeting with Maugham unarmed. Quickly apprehended, the Ngoni king was tried in a kangaroo court, sentenced to death, and executed.[58] By 6 November, 210 Ngoni villages had made contact with British authorities in Fort Liwonde and asked for permission to settle in the Upper Shire District; the token of submission they all sent in was a spear.[59]

THE REMAKING OF NGONI MILITARISM

The incorporation of the Maseko and Mpezeni's Ngoni into the British Central Africa Protectorate and North-Eastern Rhodesia, respectively, spelled the end of their war-based societies and of the raiding system that had kept inter-regimental advancement mechanisms oiled. Yet important elements of the precolonial pattern of social relationships and honor culture endured even in the changed circumstances of the early twentieth century. This is especially clear in the case of the Jere Ngoni of Mpezeni, on which the bulk of this section focuses.

The public execution of Nsingo had been explicitly used by the leaders of the occupying army to impress upon attending Ngoni dignitaries and soldiers that the days of raiding were over once and for all.[60] Raiding, however, witnessed a brief revival under the auspices of the (technically) illegal police maintained by the North Charterland Exploration Company in Fort Young in 1898–1899. Led by the notorious Ziehl, this Snider-equipped, uniformed force became the "terror of the district," and must have provided a few demobilized Ngoni fighters with the opportunity to continue to pursue their vocation. Accused of various misdemeanors—including cattle rustling in the Protectorate's territory—the NCEC police was quickly disbanded under pressures from both Robert Codrington—the BSAC's deputy administrator, who had recently relocated to Fort Jameson, the new capital of North-Eastern Rhodesia[61]—and Alfred Sharpe, the BCAP's high commissioner.[62]

Porterage in the service of the NCEC until c. 1910 and, especially, labor migration towards Southern Rhodesia—which began in earnest immediately after the defeat of 1898—offered less ephemeral arenas for the survival of significant traits of the old honor culture.[63] Lust for adventure, resilience in the face of adversity, the cultural belittlement of agricultural labor, and the readiness to put one's life at stake with a view to improving one's social standing—all these cultural elements celebrated, *inter alia*, in the imigubo war songs examined in the previous chapter might have been responsible for the rapidity with which "the idea of going to work" in the Southern Rhodesian gold mines "caught on with the Angoni" of eastern Zambia.[64] Of course, more obvious and, indeed, better-known structural determinants were also at play, for the poverty in which the Ngoni heartland was plunged by the war of 1898 was compounded in 1900 by the imposition of a hut tax, the device through which regional colonial regimes invariably sought to extract labor from their newly acquired African charges.[65] But compulsion and what historians of an earlier generation used to consider the systemic alliance between capital and the colonial state clearly do not tell the whole story of Ngoni labor migration.

The centrality of African motives and agency emerges with particular clarity when one examines the history and patterns of recruitment into colonial paramilitary police forces, the institutions that offered the most immediate opportunities for the twentieth-century remaking of Ngoni militarism.[66] Before the war of 1898, the Europeans regarded Mpezeni and his followers with a mixture of dread, revulsion, and admiration. Early in 1896, Warton, the NCEC's manager, explained that Mpezeni's "people (Zulus) have entirely driven out the original native population, and he is a terror to the surrounding villages for many miles. He is a slave-raider and exercises the powers over life and death."[67] Mpeseni—he reported after, and despite, his first positive meeting with the king—was "an absolute savage. He thinks no more of killing a man than of killing an ox, perhaps not so much."[68] Underlying these images of martial might and unrestrained brutality was the historical link between the Jere Ngoni and the Zulu of South Africa, whose military ethos and tactics, of course, had made an everlasting impression on British imperial policy makers and commentators since the 1870s. The connection was made explicit by Harry Johnston, the first consul-general and high commissioner in British Central Africa: "Mpeseni is a very powerful Chief, of Zulu or Matabele origin. . . . He can put together at least 30,000 warriors into the field, armed chiefly with spears and assegais, and attacking impetuously and with great bravery in the Zulu style."[69] Martial representations such as these, of course, could be used for alternative—but not necessarily conflicting—purposes: to convey a sense either of African inherent savagery or of the worthiness of a likely future opponent. Their sway, moreover, was generally reinforced by the war of 1898, during which—as has been seen above—some at least of the European protagonists interpreted the naked courage displayed by many Ngoni fighters in the face of the enemy's crushing technological superiority as a demonstration of lingering "Zuluness."

Images of Ngoni "pluckiness" (a favorite word in the vocabulary of "martial race" ideologues) help explain the decision to recruit some of their number into the newly formed Mashonaland Native Police (MNP), part of Southern Rhodesia's British South Africa Police (BSAP). But equally—if not more—important were the Jere Ngoni's own predilections and aspirations. Administrator Forbes, who first suggested that a Ngoni contingent be raised for service in Southern Rhodesia, stressed that the Ngoni of both Zambia and Malawi were "willing to go anywhere" and that they took "great pride in being made soldiers."[70] It was on the strength of this assessment that Colin Harding, a major in the BSAP and the future commandant of the Barotse Native Police, was dispatched to Nyasaland and then Fort Jameson with a view to enlisting some three hundred Lakeside Tonga and Ngoni police. Harding was also expected to take charge of the BSAC police in North-Eastern Rhodesia. Even though this part of the plan fell by the wayside, his recruiting mission on behalf of the MNP in the summer of 1898 was a resounding success. As soon as Harding reached Fort Jameson, between 150 and 200 Ngoni "young men" were promptly handed over to his charge by local authorities to supplement the levies he had already raised in the BCAP.[71] No coercion was apparently needed at any stage, for the Ngoni joined Harding's incipient force "readily."[72] Once in Southern Rhodesia, the Ngoni contingent from North-Eastern Rhodesia is said to have adjusted easily to paramilitary life[73]—so much so, indeed, that, upon expiry of its original term of service in 1900, some at least of its members "readily found employment as [mining] compound police."[74]

As pointed out above, by the time of Harding's visit, the exiled Mpezeni's heartland was a shadow of its former self, having seen most of its large villages and fields razed to the ground and suffered "enormous losses" in cattle.[75] To be sure, it is tempting to interpret the enthusiasm for police service reported by Harding in 1898 as a direct consequence of what the same witness called "the state of despair and desolation" then prevailing in Ungoni.[76] But—as in the case of labor migration, which kick-started at roughly the same time—an exclusive emphasis on the imperative of survival or what Timothy Stapleton calls "material motives" and "concerns" obfuscates the workings of more profound historical processes.[77] The glorified place of warfare in the Ngoni's precolonial social system, and the determination on the part of aspiring or demobilized warriors to continue to command their "right to respect" through military pursuits, were the key sociocultural forces shaping Ngoni choices. Indeed, there are some indications that, in Ngoni eyes, Harding's training camp in Fort Jameson represented the structural equivalent of an old regimental barrack, which chiefs and elders visited regularly to assess progress and which they kept well supplied with provisions.[78] This reading of Ngoni motivations, moreover, is borne out by the fact that Ungoni's gradual economic recovery in the early 1900s did not, as we shall presently see, undermine its young residents' willingness to serve in colonial forces.

For our story, North-Western Rhodesia's Barotse Native Police (BNP) is especially important, since the colonial fixation with the "trust-worthy stranger to police other strangers" meant that the Fort Jameson Ngoni were prevented from dominating the ranks of the North-Eastern Rhodesia Constabulary, the police corps whose area of operations encompassed their home region.[79] Opportunities in the CAR and, to a lesser extent, the BCAP police were also limited, mainly on account of the two forces having been built around nuclei of Lakeside Tonga and Yao volunteers before the final subjugation of the Ngoni.[80] The BNP came into existence once the cost-cutting decision was taken to replace the detachment of white Southern Rhodesian police who had been stationed in Fort Monze, North-Western Rhodesia, since 1898.[81] Officially gazetted at the beginning of 1901, the BNP was mainly employed "as a defensive force in garrisoning the different stations, patrolling their districts under Officers or white NCOs, executing warrants, [and] as escorts to District Commissioners when visiting the kraals in their District."[82] What even this dry description makes clear is that the BNP was no different from other colonial police forces, for which the prevention and detection of crime were always subordinate to the defense of the colonial order from internal threats.[83] The paramilitary character of the BNP was a direct consequence of its primary function, and it manifested itself in the heavy emphasis placed in training on arms drill and musketry practice.

Harding, the first commandant of the BNP, was from the beginning keen on employing some of the same Ngoni recruits he had earlier enrolled in the Mashonaland Native Police.[84] The plan, however, did not materialize at this early stage. By the end of 1902, the BNP consisted of 227 African noncommissioned officers and men under a dozen white commissioned and noncommissioned officers; all of the African recruits hailed from North-Western Rhodesia itself.[85] Since Harding considered this state of affairs dangerous and "unwise,"[86] the BNP ranks were supplemented with seventy-five Bemba-speakers from the neighborhood of Lake Mweru in the course of 1903.[87] But aspiring Ngoni military men were not ready to be sidelined and sought to influence recruitment patterns from below. In the early years of the century, Fort Jameson Ngoni—and even some of their confreres from the BCAP—were reported to be "constantly" travelling to Kalomo, the then capital of North-Western Rhodesia and headquarters of the BNP, "anxious to enlist in the B.N. Police, having walked from N.E. Rhodesia, on their own initiative and unaccompanied by any white man."[88]

Ngoni pressures bore the desired fruit and, in the summer of 1904, Major Carden was sent to Fort Jameson on recruiting duties. He returned to Kalomo with 124 volunteers, but, according to Harding, he could have easily enlisted as many as 500.[89] Of the 124 men brought back by Carden, as many as 90 (or 72 percent) were classified as Ngoni. Of these, the overwhelming majority came from the Fort Jameson area and, most notably, the districts under Madzimawe and Nzamane, two of the late Mpezeni's sons, who were now officially recognized as chiefs by the BSAC.[90] This large influx made the Ngoni the most heavily represented ethnic group in the BNP's rank and file. By the end of 1904, Ngoni police constituted about 27 percent of a force that now numbered 336 African NCOs and men. The second largest contingent, numbering 50 men, were Ila (14 percent).[91] Once established, Ngoni dominance of the BNP would prove long-lasting. Late in 1905, only "20 local natives" remained in the corps, the bulk of its members being "composed of natives from North Eastern Rhodesia."[92] In 1908–9, all of the new recruits required by the BNP were once more enlisted in Fort Jameson, while many of their predecessors signed up for a further term of three years.[93] By then, the Ngoni (and Bemba) had also managed to penetrate the North-Eastern Rhodesia Constabulary, where the preference for "alien native" recruits had seemingly been relaxed at the same time as it had been tightened in North-Western Rhodesia.[94]

The above discussion has some important implications. The extent to which the Jere Ngoni's own motivations and sociocultural inclinations were responsible for carving out a privileged niche for themselves in the BNP (and, earlier, in the MNP) casts doubts on approaches that foreground European agency alone in the fabrication of imperial taxonomies—including the so-called martial race ideology, which Heather Streets's work, for instance, presents as being, in essence, a one-way Victorian construction, "born out of specific recruiting needs" and nineteenth-century ideas about the supposed cultural characteristics of particular physical environment, such as the view that hot and tropical areas generated a "lazy, lascivious, passive, effeminate and degenerate . . . population. . . . Conversely, colder, more northerly climates were believed to produce and sustain hard-working, aggressive and masculine people."[95] A similarly lopsided perspective is adopted by Timothy Parsons with regards to Kenya and the King's African Rifles (KAR) in the first half of the twentieth century. As seen by Parsons, recruitment for the KAR amounted to a process in which European authorities automatically categorized as "martial" all such ethnic groups as proved willing to respond to calls for enlistment on account of their disadvantaged economic circumstances. In Parsons's understanding, then, "African societies were most 'martial' when taxation and land shortages forced them to seek paid employment, and educational limitations" reduced their options in the labor market.[96]

An alternative reading, however, has been advanced by Edward Steinhart, who emphasizes that, for all the weight of British ethnic military stereotypes and environmental fantasies, precolonial experiences were not irrelevant in shaping patterns of colonial recruitment. In the course of the nineteenth century, for instance, the Kamba of Kenya had pursued careers as elephant hunters and worked for wages in the long-distance caravan trade. Thus, behind the prominent role that they came to play as colonial soldiers and police was no doubt the fact that military life, "involving travel, discipline, and a sense of adventure, was very much consonant with Kamba cultural expectations of what was a suitable vocation for young men."[97] Military service in Ukambani, Myles Osborne argues in a similar vein, "evoked important aspects of the Kamba past relating to bravery in martial occupation, travel far from home and providing for the sustenance of the community."[98]

The case of the Ngoni—like that of the Kamba—shows that the imperial frontier was not a tabula rasa awaiting colonial inscription and that the potential for historical invention is never limitless. Nothing is made out of nothing, and contextual forces do constrain discursive ones. The colonial subjects' precontact military histories and social structures influenced their terms of engagement with European forces and, therefore, the perceptions, imaginings, and policies of the leaders of the latter. Sometimes, specific social or ethnic groups were construed as "martial"—and therefore targeted as military or paramilitary recruits—because they *were* martial. This, of course, is not to suggest the existence of a biological predisposition towards warfare among select groups. And it certainly remains the case that acknowledged "precolonial military prowess was no guarantee of martial status" under European rule.[99] The Maasai in east Africa are a good case in point.[100] But a focus on imperial interests and myth-making should not obscure the fact that the precontact trajectories and social organizations of some communities had been more deeply shaped by the experiences of warfare and militarism than those of others. Given their precolonial war-centered social system and normative universe, it is not surprising that the Ngoni came to be conceived of as Northern Rhodesia's prime "martial race." Nor is it surprising that scores of young Ngoni should have volunteered for service in paramilitary forces that appeared to hold out the promise of perpetuating such notions of honor and masculinity as had informed Ngoni sociocultural structures in the recent past.

Within this framework of continuity, however, there were also sharp ruptures. In particular—to return to the central subject of this book—the experiences of the Ngoni in colonial police forces transformed beyond recognition their understanding of firearms. Because of its paramilitary character, the BNP was equipped with single-shot Martini-Henry breech-loading rifles from the day of its inception.[101] Firearms symbolized the very essence and repressive potential of the BNP, and proficiency in their use was systematically promoted through regular musketry courses held both at the corps' headquarters and in peripheral commands.[102] In this light, a series of fairly precise parallels can be drawn between the lives and expectations of precolonial Ngoni fighters and those of BNP recruits. In precolonial times—as has been argued above and in the previous chapter—warfare had offered young Ngoni spearmen the chance to display their heroism and, by so doing, unlock the potential for self-advancement inherent in their social institutions. Given the comparatively peaceful nature of the occupation of North-Western Rhodesia by the BSAC, opportunities for real warfare while serving in the BNP were, in practice, limited to a few anti-slavery and anti-smuggling patrols (see chapter 4). Thus, rather than expressing itself in open battle and in the heat of close combat, inter- and infra-regimental competition now took the form of shooting contests, with "Marksman Badges" for proficient riflemen taking the place of the horn-made "military order" described by Wiese in the 1890s (see chapter 5).[103]

Warfare and raiding, of course, had also enabled junior Ngoni soldiers to ameliorate their socioeconomic status through the direct accumulation of both material wealth (most notably cattle) and dependents. While raiding *sensu stricto* was not openly endorsed by the BNP, police service still offered ample opportunities for preying on civilians.[104] In condemning the police's prevailing conduct, Marshall Hole, who served as North-Western Rhodesia's acting administrator in 1903, urged the BNP's commanding officers to bear in mind that the "uncivilized instincts" of local recruits could not be "eradicated" by a "few months drill and acquaintance with discipline. . . . On the other hand, the possession of firearms and their familiarity with the use of them, together with the right to wear uniform, are calculated to puff them up with a sense of their own superiority and importance, and to tempt them to adopt a bullying and overbearing demeanour when brought into contact with the unarmed population, unless they are restrained by the presence and supervision of a white Officer or Non-Commissioned Officer."[105] In emphasizing the connection between "possession of firearms" and "overbearing demeanor," Marshall Hole—his crass racism notwithstanding—was certainly pointing to a real problem—one which Regimental Order no. 1 of 23 February 1903 had sought partly to tackle by stipulating that "in future all Police going on leave will go unarmed."[106] Yet—as Acting Commandant Monro coyly put it—the tendency "to presume at times on the natives" while on active service was unaffected by the Regimental Order, and indeed it remained a distinguishing trait of the BNP for several years to come.[107]

Between September 1907 and September 1908, for instance, at least three BNP recruits were either fined or given (short) prison sentences for various sexually related offences, ranging from "assault with violence and indecent assault" (Mumbwa command) to "sending to a village and procuring by force a native woman" (Mongu) and "interfering and sleeping with a native woman bringing grain into camp" (Mumbwa).[108] Nine more policemen were punished for common or violent assault, the victims of police violence being, in two cases, "native prisoners."[109] Other crimes against civilians reported during the same period included "creating a disturbance in a native village,"

FIGURE 6.1 "Barotse Native Policeman outside hut, with kit laid out for inspection. Kalomo. 1905." Source: Robert Coryndon's Album, Livingstone Museum, Livingstone, Zambia.

"taking a basket of meal from a native woman and striking her on the head," and even "abduction."[110]

⟃

Competition between warriors for upward mobility and their exalted position vis-à-vis less militarily prepared agriculturalists had been central elements of Ngoni precolonial militarism. Both features found a new lease on life within the BNP, yet both were transformed by firearms. For Ngoni recruits—as for their precolonial predecessors—military life remained the mainspring of heroic honor, a means of masculine affirmation, and the ultimate source of socioeconomic advancement. What changed was that such desirable accomplishments were now predicated on the deployment and effective appropriation of a technology—firearms—that young Ngoni had previously rejected as inadequate to foster their self-improving goals. Within the ranks of the BNP, if not in Ungoni as a whole, the Martini-Henry had become the modern assegai.

CONCLUSION

Gun Domestication in Historical Perspective

By DRAWING on recent constructivist approaches to the history of technology and consumption, this book has attempted to offer new perspectives on the history of firearms in a large swathe of central Africa between the early nineteenth and the early twentieth century. In the process, it is hoped, a case has been made for shifting the study of technology away from its customary Western contexts and for a renewed scholarly engagement with precolonial African history in light of theoretical concepts originating from outside the field of African studies.

While sharing significant historical commonalities, the central savanna in the last century of its independence presented a politically and culturally diversified picture. This, as chapter 1 has argued, was partly a consequence of the fact that the peoples of the interior responded differently to the expansion of international trade in the nineteenth century. As the case studies discussed in the second part of the book have shown, long-established, centralizing monarchical organizations stood side by side with decentralized "stateless" societies and such more recent products of the advancing frontier of global commerce as the upper Zambezi's trading principalities and Msiri's warlord state in southern Katanga. My overarching contention—*pace* much of the existing literature—has been that imported firearms were critically important in virtually every case, but that the centrality of guns manifested itself in different ways and took different forms. The encounter between an originally exogenous technology and the variegated cultural structures and politico-economic formations that predated its introduction on the central African scene produced much heterogeneity in patterns of domestication and resulted in firearms being attributed a multiplicity of both predictable and innovative functions. In the central savanna, it has been argued throughout, guns did not work solely as hunting and military implements, and their history is poorly served by approaches that do not distinguish between the functional properties of firearms and the set of symbolic values and meanings that they were taken to encapsulate.

Beginning in the early decades of the nineteenth century, flintlock muskets imported through Angola were incorporated into the mobile hunting economies of the peoples of the headwaters of the Zambezi as critical markers of male identity and lubricants of social relationships. These understandings of guns were no less important than their deployment as tools of material production favoring a more and more intensive involvement in the harvesting of ivory and slaves for the global market. This mode of domestication was carried to its extreme consequences a few decades later in Msiri's warlord polity, where, however, largely because of its unprecedentedly exploitative character and lack of legitimacy in the eyes of its autochthonous subjects, firearms remained more closely intertwined with the prerogatives of the state and the militarized foreign elite that dominated it. In their ability to turn guns into tools of politico-economic centralization and means of military domination, Yeke rulers resembled the royal circles of Bulozi, the upper Zambezi floodplain. On the other hand, because the relationship between restored Lozi monarchs and market forces was less intimate and foundational than in the case of the Yeke, reliance on firearms did not replace the religious and ideological underpinnings of royalty in Bulozi. Rather, firearms were absorbed into royal symbolism and attributed the function of charismatic objects, serving to embody and project an exalted image of a newly rejuvenated monarchical project.

An in-depth examination of yet another sociopolitical system—that of the Ngoni of eastern Zambia and Malawi—provided the opportunity to tease out an often overlooked aspect of the agency of users, or potential users, of technology: resistance or disengagement. Characterized by limited involvement in foreign trade, these products of the large-scale violence and dislocations of the South African Mfecane appraised firearms in pejorative terms, regarding them as a threat to local concepts of honor and to the integrity and meritocratic aspects of regimental organizations. It was only the colonial conquest and the recasting of Ngoni militarism that followed it that brought about a radical reconfiguration of Ngoni understandings of firearms. Of course, it was not only in eastern Zambia and Malawi that colonialism, by transforming the overall balance of power in the central African interior, ushered in a new phase in the relations between gun technology and African societies. Again, however, historical outcomes were neither predetermined nor univocal. Thus, while gun societies withered on the upper Zambezi in the aftermath of the introduction of new gun and game laws in the 1920s, they survived in southern Katanga, where firearms, though eventually decoupled from state power, retained a more central economic and cultural position than they did elsewhere, with important consequences for the future of the region.

⤿

Do the findings of this book and the culturally sensitive approach that made them possible speak to pressing contemporary concerns? My bold final contention is that they do—and nowhere more so than in the conflict zones of the eastern Congo. Lying close to the area we have dealt with in this book, and having been affected by related historical developments in the nineteenth century, this troubled region represents a well-known instance of ingrained strife and gun proliferation, and of the intractable problems involved in bringing both phenomena under control.[1]

Careful students of the crisis that continues to plague the Kivus, Manyema, and northern Katanga even after the formal cessation of open hostilities between state actors in the early 2000s have highlighted the local roots of fighting and militia formation. In soliciting international peace-builders to sideline macro-level and "top-down" explanatory frameworks, both Koen Vlassenroot and Séverine Autesserre, for instance, have stressed the urgent need to form a better understanding of the fighters' agendas and motives. Critically important, in this reading, is the extent to which participation in militarized networks offers a tangible "alternative to exclusion and a way to benefit from modernisation" to scores of "young men who were marginalised in customary networks of dependency," land apportionment mechanisms, and educational opportunities.[2] In restructuring their members' identities, Mayi-Mayi formations—very much like similar armed groups elsewhere in modern Africa[3]—draw on inextricably hybridized endogenous and exogenous symbols. Thus, in what Christopher Bayly would call the "bodily practice" of militiamen,[4] medicinal tattoos and magically treated water (from which the militias derive their general name) stand side by side with such (commonly looted) imported goods as "sun glasses, mobile phones, smart shoes and clothes, motor cars, motorcycles, etc." In an "economy of scarcity," the "symbolic value of these objects" as "fetishes of modernity" can scarcely be overemphasized.[5] But, of course, it is the ever readily available Kalashnikov—ubiquitous marker of modernity and masculinity in the eyes of its users and exclusive "tool of promotion in a society that [could] not integrate them any more"—that most aptly embodies the ethos of Mayi-Mayi grassroots fighters.[6] Omari "*Double-Lame,*" the *kadogo*, or child soldier, imagined by In Koli Jean Bofane, had been instructed by his minders to have the greatest of respect for the AK-47 that he had been immediately entrusted with upon joining an unspecified rebel group in the east. Expected to take care of the rifle "as the apple of his eye," he learned to disassemble and reassemble it. So familiar had he become with the functioning of the weapon that he ended up "being at one with it." Later on, it was the disposal of his gun in the Gombe Stream that symbolized Omari's desertion from the newly integrated army and his acceptance into the ranks of the street kids roaming Kinshasa's Grand Marché.[7]

Yet, with few exceptions, both modern scholars and international organizations and NGOs tend to ignore, or at least make very little of, the fact that the eastern Congo has been there before, as it were. In the late nineteenth century, as the region was incorporated into Arab-Swahili and Sudanese commercial networks, large parts of it were brought under the sway of the predatory trading "empire" revolving around the person of the Zanzibari merchant-warrior Tippu Tip (the beginnings of whose conquering trajectory between Lakes Tanganyika and Mweru have been touched upon in chapter 1). New commercially oriented polities under local warlords—northern reverberations of the same processes we have seen at work in southern Katanga in the second half of the century (chapter 3)—also mushroomed at this time, either in competitive or in collaborative relation with Tippu Tip's regime. In this volatile context, characterized by widespread social breakdown and an atmosphere of pervasive threat, new opportunities for social mobility, material gain, and identity formation were thrown open to uprooted young men, known in Swahili as *ruga ruga*, who in many ways were both the consequences and the causes of the dislocations of the slave and ivory trades. For instance, besides *waungwana* (Islamized, Swahili-speaking retainers), Nyamwezi porters, and armed slaves, the "motley," flintlock-wielding force that Tippu Tip made available to Stanley in 1876 also included "about fifty youths." These precursors of today's Congolese kadogo ranged "from ten to eighteen years of age, being trained by Tippu-Tib as gun-bearers, house servants, scouts, cooks, carpenters, house-builders, blacksmiths, and leaders of trading parties. . . . Such young fellows are useful to him; they are more trustworthy than adults, because they look up to him as their father; and know that if they left him they would inevitably be captured by a less humane man."[8] Two years earlier, reporting from Kasongo, one of the two Arab-Swahili capitals in Manyema in the 1870s, Cameron had been surprised to discover that one unnamed Arab trader employed as many as "six hundred Nyamwezi, all armed with guns." "These fellows"—he explained—"get little or no pay, but are allowed to loot the country all round in search of subsistence and slaves. Some of the slaves they keep for themselves, giving their employers a sufficient number in return for the powder supplied to enable them to oppress the natives."[9]

What was true of the militarized followers of Tippu Tip and other coastal traders was also true of the gunmen at the service of local political opportunists, such as the Songye warlord Lumpungu, in present-day Kasaï-Oriental, on the western borders of the Arab-Swahili sphere of influence. In 1887, Hermann von Wissmann, who had already met Lumpungu five years earlier, learned that, having traded much with the Arab-Swahili of Manyema in the past, his host had now fallen out with his erstwhile backers. Wissmann found Lumpungu "changed from his former modest and amiable deportment: his constant persecution during the last few years had made him restless and savage in a way that he had not been formerly." Still, both he and the "many thousands" of boisterous, "warlike" fighters who surrounded him, surviving on a diet of palm wine and flaunting their percussion-lock muskets, presented "a great contrast" to Wissmann's emaciated followers. Controlling large food supplies despite residing in a "district of starvation," Lumpungu's men persuaded the members of Wissmann's caravan to part with most of their guns, powder, and percussion caps; the German traveler was dismayed at the prospect of hostilities "breaking out among the mostly drunken hosts of warriors."[10]

It was in the late nineteenth century, then, that the eastern Congo first witnessed the emergence of those whom Reid, with reference to northern Tanzania, describes as "rootless, displaced communities, simultaneously traumatized and criminalized, which lived off the violence of the slave trade."[11] This, too, was the moment that mass violence first became a defining feature of life in the region and that African societies and gun technology were brought into intimate contact with one another. The practical and expressive meanings attributed to firearms at the time—means of commodity extraction and personal enrichment, of course, but also means of community protection against external plunder and symbols of youthful machismo—are still dominant today, no doubt also because the brutality of the colonial conquest and early exploitation of the area and the large-scale turmoil associated with the period of decolonization and the coeval Simba rebellion of the mid-1960s did nothing to lessen the cogency of such patterns of gun domestication. This, of course, is not to argue that nothing has changed in the eastern Congo. That would be both absurd and unhelpful. But, surely, ongoing peacekeeping and disarmament efforts stand a better chance of succeeding once it is realized that militias and their firearms are not modern-day aberrations, but the latest epiphenomena of politics of predation that date back to the region's incorporation into the world economy in the second half of the nineteenth century and the related rise of warlordism. A thick layer of past technological engagements and historical understandings of military-civilian relations must be peeled off before such efforts can be expected to bear any enduring fruits.

Similar considerations should perhaps be brought to bear with regards to other twentieth-first-century African trouble spots, such as the North Rift Valley, where the "weaponization of society" (to use Kennedy Mkutu's expression) began much earlier than is commonly realized by state and development agencies,[12] and, further north, the border between Ethiopia and South Sudan, where the broader symbolic significance of rifles, and their entanglements in "local concepts of male potency, beauty and strength," may have been overlooked as a result of an exclusive concentration on their obvious service functions—that is, on their deployment in inter-communal feuding and large-scale warfare.[13] But the failure to foreground the complex social relationships revolving around guns is not just the consequence of the adoption of narrowly politico-economic perspectives, for such myopia is also characteristic of the vast literature devoted to the religious aspects of postcolonial armed conflicts.[14] It is indeed a peculiar paradox that even such culturally inflected studies should have directed so little attention to the symbolic significance of the technology that has made so many of the conflicts they examine so deadly. Yet scholarly disinterest has clearly not prevented Africans from continuing to debate the meaning of firearms. In Mozambique, for instance, K. B. Wilson tells us in a tantalizing aside, early adherents to the government-supported Naprama movement in 1990 shunned firearms in their confrontations with Renamo, since the "cold steel" of the spears they fought with was construed as being more "moral" than their oppressors' weapons.[15] The point hardly needs to be belabored that a full appreciation of hostilities in late twentieth-century Mozambique and elsewhere must make room for this and other local understandings of the hardware of violence at the disposal of belligerents.

This brief excursus into eastern Congo and current affairs lends further weight to Richard Reid's contention that precolonial history remains critical to understanding contemporary Africa and that it behooves historians of the continent to seek to reconnect its deep past with the challenges of the present.[16] True in most fields of scholarly endeavor, the claim that a grounding in precolonial history "facilitates a sharper appreciation of more recent events" also applies to the sphere of technology and its social construction.[17] Firearms are not, or not only, a "problem." They are also a technological artifact with which African societies in eastern Congo and elsewhere in the central African interior have now engaged for between 150 and 200 years. In the process, the same item was successively re-innovated, with indigenous new meanings and functions being conferred to it, over and above the intentions of its Western producers. Speaking to a troubled modernity, these meanings and functions account for the continuing centrality of firearms to social life in many parts of the continent. Policy-makers and international interveners ignore them at their peril—and to the detriment of the people they strive, but often fail, to assist.

Notes

INTRODUCTION: FIREARMS AND THE HISTORY OF TECHNOLOGY IN AFRICA

1. The quoted expression is Daniel Roche's, as cited by Leor Halevi, in Leora Auslander et al., "AHR Conversation: Historians and the Study of Material Culture," *American Historical Review* 114, no. 5 (2009): 1365.

2. Honoré de Balzac, "Another Study of Womankind" (1842), in Honoré de Balzac, *The Human Comedy: Selected Stories* (New York: New York Review Books, 2014), 34–35.

3. Richard J. Reid, "Past and Presentism: The 'Precolonial' and the Foreshortening of African History," *Journal of African History* 52, no. 2 (2011): 135–55.

4. David Edgerton, "Innovation, Technology, or History? What is the Historiography of Technology About?," *Technology and Culture* 51, no. 3 (2010): 683.

5. Trevor J. Pinch and Wiebe E. Bijker, "The Social Construction of Facts and Artifacts: Or How the Sociology of Science and the Sociology of Technology Might Benefit Each Other," *Social Studies of Science* 14, no. 3 (1984): 399–441. This was reprinted in *The Social Construction of Technological Systems: New Directions in the Sociology and History of Technology*, ed. Wiebe E. Bijker, Thomas P. Hughes, and Trevor J. Pinch (Cambridge, MA: MIT Press, 1987).

6. Wiebe E. Bijker and John Law, "General Introduction," in *Shaping Technology/Building Society: Studies in Sociotechnical Change*, ed. Wiebe E. Bijker and John Law (Cambridge, MA: MIT Press, 1992), 4.

7. Ibid., 8.

8. Ronald Kline and Trevor J. Pinch, "Users as Agents of Technological Change: The Social Construction of the Automobile in the Rural United States," *Technology and Culture* 37, no. 4 (1996): 767.

9. Gabrielle Hecht, *The Radiance of France: Nuclear Power and National Identity after World War II* (Cambridge, MA: MIT Press, 1998), 9.

10. Pinch and Bijker, "Social Construction," 428.

11. Hughie Mackay and Gareth Gillespie, "Extending the Social Shaping of Technology Approach: Ideology and Appropriation," *Social Studies of Science* 22, no. 4 (1992): 685–716.

12. For a lucid introduction, see Dan Hicks, "The Material-Cultural Turn: Event and Effect," in *The Oxford Handbook of Material Culture Studies*, ed. Dan Hicks and Mary C. Beaudry (Oxford: Oxford University Press, 2010), 25–98.

13. Nelly E. J. Oudshoorn and Trevor J. Pinch, "Introduction: How Users and Non-Users Matter," in *How Users Matter: The Co-Construction of Users and Technology*, ed. Nelly E. J. Oudshoorn and Trevor J. Pinch (Cambridge, MA: MIT Press, 2005), 12.

14. Nicholas Thomas, *Entangled Objects: Exchange, Material Culture, and Colonialism in the Pacific* (Cambridge, MA: Harvard University Press, 1991), 125. And see, of course, Arjun Appadurai, ed., *The Social Life of Things: Commodities in Cultural Perspective* (Cambridge: Cambridge University Press, 1986).

15. Grant D. McCracken, *Culture and Consumption: New Approaches to the Symbolic Character of Consumer Goods and Activities* (Bloomington: Indiana University Press, 1988), xi. See also Daniel Miller, *Material Culture and Mass Consumption* (Oxford: Blackwell, 1987), and, by the same author, "Consumption and Commodities," *Annual Review of Anthropology* 24 (1995): esp. 146–47.

16. See, e.g., Paula Findlen, ed., *Early Modern Things: Objects and Their Histories, 1500–1800* (London: Routledge, 2013). The quoted expression comes from Daniel Miller and Christopher Tilley, "Editorial," *Journal of Material Culture* 1, no. 1 (1996): 11.

17. Deborah Cohen, *Household Gods: The British and Their Possessions* (New Haven: Yale University Press, 2006).

18. Anne S. Laegran, "Escape Vehicles? The Internet and the Automobile in a Local-Global Intersection," in Oudshoorn and Pinch, *How Users Matter*, 82; see also Merete Lie and Knut H. Sørensen, ed., *Making Technology Our Own? Domesticating Technology into Everyday Life* (Oslo: Scandinavian University Press, 1996).

19. Pinch and Bijker, "Social Construction," 419.

20. Jeremy Prestholdt, *Domesticating the World: African Consumerism and the Genealogies of Globalization* (Berkeley: University of California Press, 2008).

21. David Howes, "Introduction: Commodities and Cultural Borders," in *Cross-Cultural Consumption: Global Markets, Local Realities*, ed. David Howes (London: Routledge, 1996), 5–6.

22. Marshall Sahlins, *Islands of History* (Chicago: University of Chicago Press, 1985), vii.

23. Domestication/creolization processes, of course, are not confined to material objects, but also encompass ideas, images, and media forms. See, for instance, Daniel Miller's analysis of the way in which the content of an imported American soap opera was received and reconstituted in Trinidad, becoming a "key instrument for forging a highly specific sense of Trinidadian culture." Daniel Miller, "The Young and the Restless in Trinidad: A Case of the Local and the Global in Mass Consumption," in *Consuming Technologies: Media and Information in Domestic Spaces*, ed. Roger Silverstone and Eric Hirsch (London: Routledge, 1992), 165.

24. Ralph A. Austen and Daniel R. Headrick, "The Role of Technology in the African Past," *African Studies Review* 26, nos. 3–4 (1983): 165.

25. Daniel R. Headrick, *The Tools of Empire: Technology and European Imperialism in the Nineteenth Century* (Oxford: Oxford University Press, 1981).

26. Headrick's influence is such that it even impinges upon mainstream commentators on the same subjects. See, for instance, Ian Jack, "Four historians, two arguments, nobody dead. Does it matter? Well, yes," *The Guardian*, 19 November 2011.

27. Daniel R. Headrick, *Power over Peoples: Technology, Environments, and Western Imperialism, 1400 to the Present* (Princeton: Princeton University Press, 2010).

28. See, for instance, the stimulating Mirjam de Bruijn, Francis Nyamnjoh, and Inge Brinkman, ed., *Mobile Phones: The New Talking Drums of Everyday Africa* (Bamenda, Cameroon: Langaa RPCIG, 2009).

29. Brian Larkin, *Signal and Noise: Media, Infrastructure, and Urban Culture in Nigeria* (Durham: Duke University Press, 2008); Jan-Bart Gewald, Sabine Luning, and Klaas van Walraven, ed., *The Speed of Change: Motor Vehicles and People in Africa, 1890–2000* (Leiden: Brill, 2009).

30. David Arnold, "Europe, Technology, and Colonialism in the 20th Century," *History and Technology* 21, no. 1 (2005): 97. David Arnold and Erich DeWald, "Cycles of Empowerment? The Bicycle and Everyday Technology in Colonial India and Vietnam," *Comparative Studies in Society and History* 53, no. 4 (2011): 971–96, offers an excellent template for how to go about rectifying this unsatisfactory situation.

31. Howes, "Introduction," 3.

32. Timothy Burke, *Lifebuoy Men, Lux Women: Commodification, Consumption, and Cleanliness in Modern Zimbabwe* (Durham: Duke University Press, 1996), 10.

33. Prestholdt, *Domesticating the World*, 93.

34. I borrow the quoted expression from Mackay and Gillespie, "Extending the Social Shaping of Technology Approach," 704.

35. Dmitri van den Bersselaar, *The King of Drinks: Schnapps Gin from Modernity to Tradition* (Leiden: Brill, 2007).

36. Ibid., 33.

37. On which, see also Emmanuel Akyeampong, *Drink, Power, and Cultural Change: A Social History of Alcohol in Ghana, c. 1800 to Recent Times* (Portsmouth, NH: Heinemann, 1996), 9, 28–29.

38. van den Bersselaar, *King of Drinks*, 150. For the history of alcohol in east Africa, see Justin Willis, *Potent Brews: A Social History of Alcohol in East Africa, 1850–1999* (Oxford: James Currey, 2002).

39. Jean Comaroff, "The Empire's Old Clothes: Fashioning the Colonial Subject," in Howes, *Cross-Cultural Consumption*, 26–31. For a fuller version, see John L. Comaroff and Jean Comaroff, *Of Revelation and Revolution*, vol. 2, *The Dialectics of Modernity on a South African Frontier* (Chicago: University of Chicago Press, 1997), chap. 5.

40. Jean Allman, "Fashioning Africa: Power and the Politics of Dress," in *Fashioning Africa: Power and the Politics of Dress*, ed. Jean Allman (Bloomington: Indiana University Press, 2004), 6.

41. The phrase comes from Richard J. Reid, *War in Pre-Colonial Eastern Africa: The Patterns and Meanings of State-Level Conflict in the Nineteenth*

Century (Oxford: James Currey, 2007), 11. See also John Lamphear, "Introduction," in *African Military History*, ed. John Lamphear (Aldershot: Ashgate, 2007), xix.

42. See, e.g., Shula Marks and Anthony Atmore, "Firearms in Southern Africa: A Survey"; Anthony Atmore, J. M. Chirenje, and S. I. Mudenge, "Firearms in South Central Africa"; and Jeff J. Guy, "A Note on Firearms in the Zulu Kingdom with Special Reference to the Anglo-Zulu War, 1879," all in *Journal of African History* 12, no. 4 (1971).

43. *Journal of African History* 12, nos. 2 and 4 (1971). To these one must add a coeval essay of special relevance to the region with which this book is concerned: Andrew D. Roberts, "Firearms in North-Eastern Zambia before 1900," *Transafrican Journal of History* 1, no. 2 (1971): 3–21.

44. William K. Storey, *Guns, Race, and Power in Colonial South Africa* (Cambridge: Cambridge University Press, 2008), 6.

45. Rory Pilossof, "'Guns Don't Colonise People . . .': The Role and Use of Firearms in Pre-colonial and Colonial Africa," *Kronos* 36, no. 1 (2010): 266–77.

46. John K. Thornton, *Warfare in Atlantic Africa, 1500–1800* (London: UCL Press, 1999); Richard J. Reid, *War in Pre-colonial Eastern Africa*, and, by the same author, *Warfare in African History* (Cambridge: Cambridge University Press, 2012); Edward I. Steinhart, *Black Poachers, White Hunters: A Social History of Hunting in Colonial Kenya* (Oxford: James Currey, 2006).

47. Reid, *War in Pre-colonial Eastern Africa*, 41. See also his discussion at 13, and in *Warfare in African History*, 108.

48. Storey, *Guns, Race, and Power*, 12–13.

49. Ibid., 258; see also 246–47.

50. Jeff Ramsey, "Firearms in Nineteenth-Century Botswana: The Case of Livingstone's 8-Bore Bullet," *South African Historical Journal* 66, no. 3 (2014): 442.

51. Clapperton C. Mavhunga, "The Mobile Workshop: Mobility, Technology, and Human-Animal Interaction in Gonarezhou (National Park), 1850–present" (PhD diss., University of Michigan, 2008), iii, 1.

52. Ibid., chap. 1.

53. Ibid., 102.

54. Clapperton C. Mavhunga, "Firearms Diffusion, Exotic and Indigenous Knowledge Systems in the Lowveld Frontier, South Eastern Zimbabwe, 1870–1920," *Comparative Technology Transfer and Society* 1, no. 2 (2003): 201–31.

55. Arnold and DeWald, "Cycles of Empowerment?," 973.

56. Oudshoorn and Pinch, "Introduction," 17.

57. Jonathan A. Grant, *Rulers, Guns, and Money: The Global Arms Trade in the Age of Imperialism* (Cambridge, MA: Harvard University Press, 2007); Emrys Chew, *Arming the Periphery: The Arms Trade in the Indian Ocean during the Age of Global Empire* (New York: Palgrave Macmillan, 2012).

58. See, e.g., Joseph C. Miller, "Imports at Luanda, Angola: 1785–1823," in *Figuring African Trade: Proceedings of the Symposium on the Quantification and Structure of the Import and Export and Long-Distance Trade of Africa in*

the *Nineteenth Century (c. 1800–1913)*, ed. G. Liesegang, H. Pasch, and A. Jones (Berlin: Dietrich Reimer Verlag, 1986), 176, 191–92.

59. Arnold and DeWald, "Cycles of Empowerment?," 995.

60. David M. Gordon, *Invisible Agents: Spirits in a Central African History* (Athens: Ohio University Press, 2012), 6, 10. Gordon follows in the footsteps of Stephen Ellis and Gerrie ter Haar, *Worlds of Power: Religious Thought and Political Practice in Africa* (New York: Oxford University Press, 2004).

61. Joseph C. Miller, "Cokwe Trade and Conquest in the Nineteenth Century," in *Pre-Colonial African Trade: Essays on Trade in Central and Eastern Africa before 1900*, ed. Richard Gray and David Birmingham (London: Oxford University Press, 1970), 176; Jean-Luc Vellut, "L'économie internationale des côtes de Guinée Inférieure au XIXe siècle," 170, and Maria Emilia Madeira Santos, "Tecnologias em presença: Manufacturas Europeias e artefactos Africanos (c. 1850–1880)," 221, both in *Actas de I reunião internacional de história de Africa: Relação Europa-África no 3° quartel do Séc. XIX*, ed. Maria Emilia Madeira Santos (Lisbon: Instituto de Investigação Científica Tropical, 1989). For a famous West African example, see Martin Legassick, "Firearms, Horses and Samorian Army Organization 1870–1898," *Journal of African History* 7, no. 1 (1966): 95–115.

62. Mavhunga, "Mobile Workshop," 74; Gerald M. Berg, "The Sacred Musket: Tactics, Technology, and Power in Eighteenth-Century Madagascar," *Comparative Studies in Society and History* 27, no. 2 (1985): 275–76.

63. Nicole Boivin, *Material Cultures, Material Minds: The Impact of Things on Human Thought, Society, and Evolution* (Cambridge: Cambridge University Press, 2010), 129–30; Alfred Gell, *Art and Agency: An Anthropological Theory* (Oxford: Oxford University Press, 1998).

64. Thomas J. Misa, quoted in Boivin, *Material Cultures*, 161.

65. Ann B. Stahl, "Material Histories," in *The Oxford Handbook of Material Culture Studies*, ed. Dan Hicks and Mary C. Beaudry (Oxford: Oxford University Press, 2010), 153–54.

66. Hicks, "Material-Cultural Turn," 84.

67. See, e.g., John M. MacKenzie, *The Empire of Nature: Hunting, Conservation and British Imperialism* (Manchester: Manchester University Press, 1988); and Steinhart, *Black Poachers, White Hunters*.

68. Storey, *Guns, Race, and Power*; Steinhart, *Black Poachers, White Hunters*. See also Keith Shear, "'Taken as Boys': The Politics of Black Police Employment and Experience in Early Twentieth-Century South Africa," in *Men and Masculinities in Modern Africa*, ed. Lisa A. Lindsay and Stephan F. Miescher (Portsmouth, NH: Heinemann, 2003), 109–27; Bill Nasson, "'Give Him a Gun, NOW': Soldiers but Not Quite Soldiers in South Africa's Second World War, 1939–1945," in *A Cultural History of Firearms in the Age of Empire*, ed. Karen Jones, Giacomo Macola, and David Welch (Farnham: Ashgate, 2013), 191–210; and Julie Bonello, "The Development of Early Settler Identity in Southern Rhodesia: 1890–1914," *International Journal of African Historical Studies* 43, no. 2 (2010): esp. 350–51. For a West African example,

see Lynn Schler, "Bridewealth, Guns and Other Status Symbols: Immigration and Consumption in Colonial Duala," *Journal of African Cultural Studies* 16, no. 2 (2003): esp. 229–31.

69. O. Sibum, in Auslander et al., "AHR Conversation," 1384.

70. Here, I am adopting a slightly modified version of the geographical framework of Thomas Q. Reefe, "The Societies of the Eastern Savanna," in *History of Central Africa*, vol. 1, ed. David Birmingham and Phyllis M. Martin (London: Longman, 1983), 160–204, on which this and the next two paragraphs draw heavily.

71. Ibid., 166.

72. Reid, *Warfare*, esp. chap. 5.

73. The quoted expression is used by both Roberts, "Firearms in North-Eastern Zambia before 1900," 16, and Steven Feierman, *The Shambaa Kingdom: A History* (Madison: University of Wisconsin Press, 1974), 142.

74. Isabel de Castro Henriques sought to solve the problem in a different way, introducing a rigid distinction between groups such as the Imbangala, in which firearms served, alongside other imported products, as "luxury goods that confirmed the chiefs' power and status," and communities such as the Chokwe, in which firearms entered the productive process as "tools of production." The evidence pertaining to the gun societies examined in the second part of this book, however, strongly suggests that Henriques' typology does not exhaust the range of possible African understandings of firearms. Isabel de Castro Henriques, "Armas de fogo em Angola no século XIX: Uma interpretação," in Santos, *Actas de I reunião internacional de história de Africa*, 418, 422.

75. See, e.g., Tim Youngs, *Travellers in Africa: British Travelogues, 1850–1900* (Manchester: Manchester University Press, 1994), and Mary Louise Pratt, *Imperial Eyes: Travel Writing and Transculturation* (London: Routledge, 1992).

76. Richard J. Reid, "Revisiting Primitive War: Perceptions of Violence and Race in History," *War and Society* 26, no. 2 (2007): 1. See also Lamphear, "Introduction," xi.

77. Earlier sources might, in this respect at least, be more satisfying. See, for instance, the detailed discussion of early modern African military tactics offered by John K. Thornton, "The Art of War in Angola, 1575–1680," *Comparative Studies in Society and History* 30, no. 2 (1998): 360–78.

78. Roy C. Bridges, "Explorers' Texts and the Problem of Reactions by Non-Literate Peoples: Some Nineteenth-Century East African Examples," *Studies in Travel Writing* 2, no. 1 (1998): 69.

79. Richard J. Reid, "Violence and Its Sources: European Witnesses to the Military Revolution in Nineteenth-Century Eastern Africa," in *The Power of Doubt: Essays in Honor of David Henige*, ed. Paul S. Landau (Madison: Parallel Press, University of Wisconsin Libraries, 2011), 56. For an earlier defense of the value of nineteenth-century travelogues as sources, see Andrew D. Roberts, "Livingstone's Value to the Historian of African Societies," in *David*

Livingstone and Africa, ed. Centre of African Studies (Edinburgh: University of Edinburgh, Centre of African Studies, 1973), esp. 56–60.

80. Johannes Fabian, *Out of Our Minds: Reason and Madness in the Exploration of Central Africa* (Berkeley: University of California Press, 2000), chap. 2; Beatrix Heintze, "Hidden Transfers: Luso-Africans as European Explorers' Experts in Nineteenth-Century West-Central Africa," in *The Power of Doubt: Essays in Honor of David Henige*, ed. Paul S. Landau (Madison: Parallel Press, University of Wisconsin Libraries, 2011), 19–40; and Mathilde Leduc-Grimaldi, "'This way!' Aperçu des apports africains aux expéditions européennes du XIXe siècle: Porteurs, éclaireurs et interprètes," in *L'Afrique belge aux XIXe et XXe siècles: Nouvelles recherches et perspectives en histoire coloniale*, ed. Patricia Van Schuylenbergh, Catherine Lanneau, and Pierre-Luc Plasman (Brussels: PIE–Peter Lang, 2014), 89–99.

81. Reid, "Violence and Its Sources," 45.

82. Roy C. Bridges, "Nineteenth-Century East African Travel Records," in "European Sources for Sub-Saharan Africa before 1900: Use and Abuse," ed. Beatrix Heintze and Adam Jones, special issue, *Paideuma* 33 (1987): 179.

83. William G. Clarence-Smith, "A Note on the 'Ecole des *Annales*' and the Historiography of Africa," *History in Africa* 4 (1977): 279. The main published compilations of political oral traditions relating to this book's study area are: Adolphe D. Jalla, *Litaba za Sichaba sa ma-Lozi* (1910; repr., Dundee, South Africa: Ebenezer Press, 1922); Mukanda Bantu, "Les mémoires de Mukande Bantu," *Bulletin de la Société Belge d'Études Coloniales* 27, nos. 5–6 (1919): 251–77, nos. 9–10 (1919): 497–521; T. Cullen Young, *Notes on the History of the Tumbuka-Kamanga Peoples in the Northern Province of Nyasaland* (1932; repr., London: Frank Cass, 1970); Jan van Sambeek et al., *Ifya Bukaya: Third Bemba Reader* (Chilubula, Zambia: White Fathers, 1932), and *Ifya Bukaya: Fourth Bemba Reader* (Chilubula, Zambia: White Fathers, n.d.); J. P. Bruwer, *Mbiri ya Angoni* (Mkhoma, Malawi: DRC Mission Press, 1941); Yesaya M. Chibambo, *My Ngoni of Nyasaland*, trans. Charles Stuart (London: United Society for Christian Literature, n.d. [1942]); Simon Jilundu Chibanza, *Kaonde History*, part 2 of J. T. Munday and Simon Jilundu Chibanza, *Central Bantu Historical Texts*, vol. 1 (Lusaka: Rhodes-Livingstone Institute, 1961); Mwata Kazembe XIV [Shadreck Chinyanta Nankula], *Central Bantu Historical Texts*, vol. 2, *Historical Traditions of the Eastern Lunda*, trans. Ian G. Cunnison (Lusaka: Rhodes-Livingstone Institute, 1962); Ian Linden, "Some Oral Traditions from the Maseko Ngoni," *Society of Malawi Journal* 24, no. 2 (1971): 60–73; Mose K. Sangambo, *The History of the Luvale People and Their Chieftainship*, 2nd ed. (Zambezi, Zambia: Mize Palace, n.d. [1985?]). Fuller bibliographical listings are available in Giacomo Macola, "Historical and Ethnographical Publications in the Vernaculars of Colonial Zambia: Missionary Contribution to the 'Creation of Tribalism,'" *Journal of Religion in Africa* 33, no. 4 (2003): 343–64.

84. Jan Vansina, "For Oral Tradition (But Not against Braudel)," *History in Africa* 5 (1978): 354.

85. On "feedback," see David Henige, "Truths Yet Unborn? Oral Tradition as a Casualty of Culture Contact," *Journal of African History* 23, no. 3 (1982): 395–412.

86. For a recent example, see Allen F. Isaacman and Barbara S. Isaacman, *Slavery and Beyond: The Making of Men and Chikunda Ethnic Identities in the Unstable World of South-Central Africa, 1750–1920* (Portsmouth, NH: Heinemann, 2004), esp. 146–47, 219–20.

87. Especially useful published collections are: Margaret Read, "Songs of the Ngoni People," *Bantu Studies* 11, no. 1 (1937): 1–35; Antoine Mwenda Munongo, ed. and trans., "Chants historiques des Bayeke: Recueillis à Bunkeya et ailleurs," *Problèmes Sociaux Congolais* 77 (1967): 35–139; Léon Verbeek, *L'histoire dans les chants et les danses populaires: La zone culturelle Bemba du Haut-Shaba (Zaïre)*, Enquêtes et documents d'histoire Africaine 10 (Louvain-la-Neuve, Belgium: Centre d'histoire de l'Afrique, 1992); Léon Verbeek, *Le chasseur africain et son monde: Chansons de chasse du sud-est du Katanga* (Tervuren, Belgium: Musée royal de l'Afrique centrale, 2007), http://www.africamuseum.be/research/publications/rmca/online/chansons%20de%20chasse.pdf.

CHAPTER 1: POWER AND INTERNATIONAL TRADE IN THE SAVANNA

1. Steven Feierman, "The Shambaa," and Andrew D. Roberts, "The Nyamwezi," both in *Tanzania before 1900*, ed. Andrew D. Roberts (Nairobi: East African Publishing House, 1968), 1–15 and 117–50; Steven Feierman, *The Shambaa Kingdom: A History* (Madison: University of Wisconsin Press, 1974), esp. chaps. 5–7. For more recent discussions, see Richard J. Reid, *Warfare in African History* (Cambridge: Cambridge University Press, 2012), chap. 5, and Michael Pesek, "*Ruga-ruga*: The History of an African Profession, 1820–1918," in *German Colonialism Revisited: African, Asian, and Oceanic Experiences*, ed. Nina Berman, Klaus Mühlhahn, and Patrice Nganang (Ann Arbor: University of Michigan Press, 2014), 85–100.

2. The two adjectives come from Andrew D. Roberts, "Firearms in North-Eastern Zambia before 1900," *Transafrican Journal of History* 1, no. 2 (1971): 16–17; the broader point from John Iliffe, *A Modern History of Tanganyika* (Cambridge: Cambridge University Press, 1979), 53–66.

3. I borrow this captivating expression from Thomas Q. Reefe, "The Societies of the Eastern Savanna," in *History of Central Africa*, vol. 1, ed. David Birmingham and Phyllis M. Martin (London: Longman, 1983), 160.

4. Andrew D. Roberts, *A History of Zambia* (London: Heinemann, 1976), 78. My understanding of the concept of "state" in the context of the central African savanna owes much to the old, but still brilliant, discussion provided by Joseph C. Miller, *Kings and Kinsmen: Early Mbundu States in Angola* (Oxford: Clarendon, 1976), esp. 266–70.

5. John Darwin, *After Tamerlane: The Rise and Fall of Global Empires, 1400–2000* (London: Penguin, 2008), 314.

6. Pierre de Maret, "Sanga: New Excavations, More Data and Some Related Problems," *Journal of African History* 18, no. 3 (1977): esp. 324–25, 334.

7. Thomas Q. Reefe, *The Rainbow and the Kings: A History of the Luba Empire to 1891* (Berkeley: University of California Press, 1981), 67–72; and Reefe, "Societies," 162–63.

8. Reefe, *The Rainbow and the Kings*, 84.

9. Ibid., 58–62.

10. For reasons that will become clearer over the course of the next few pages, I follow Jeff Hoover's lead and employ the ethnonym "Ruund" to designate the central Lunda and their political creation: the kingdom of the *Mwant Yavs* on the upper Mbuji-Mayi River. "Lunda," on the other hand, refers to the broader sphere influenced by—but politically independent from—the Ruund. (J. Jeffrey Hoover, "The Seduction of Ruwej: Reconstructing Ruund History [The Nuclear Lunda: Zaire, Angola, Zambia]," 2 vols. [PhD diss., Yale University, 1978]). Another caveat is in order. In dating the beginning of the *Mwant Yav* dynasty to c. 1700, I deliberately choose the least controversial possible estimation. Ideas about the timing of the emergence of the Ruund state have been shaped by different readings of their traditions of origin. The debate is important, but a little too erudite for my present purposes. Interested readers should consult Jean-Luc Vellut, "Notes sur le Lunda et la frontière luso-africaine (1700–1900)," *Études d'Histoire Africaine* 3 (1972): 64–69; Joseph C. Miller, "The Imbangala and the Chronology of Early Central African History," *Journal of African History* 13, no. 4 (1972): 549–74; Hoover, "Seduction of Ruweji," vol. 1, chap. 5; and Jan Vansina, "It Never Happened: Kinguri's Exodus and Its Consequences," *History in Africa* 25 (1998): 387–403.

11. Eva Sebestyen and Jan Vansina, ed. and trans., "Angola's Eastern Hinterland in the 1750s: A Text Edition and Translation of Manoel Correia Leitão's 'Voyage' (1755–1756)," *History in Africa* 26 (1999): 347.

12. Hoover, "Seduction of Ruweji," vol. 1, chaps. 5–6; Reefe, *The Rainbow and the Kings*, 77.

13. With positional succession, the successor to a name-title, be he (or she) an heir or the appointee of a higher authority, acquired his predecessor's identity, responsibilities, and wives and children. Perpetual kinship was a logical development, resulting in the successor also taking up the kinship ties of his predecessor.

14. Jan Vansina, *How Societies Are Born: Governance in West Central Africa before 1600* (Charlottesville: University of Virginia Press, 2004), 256.

15. Roberts, *History of Zambia*, 85.

16. Ibid., 86. For a fuller treatment of the tenure and ritual rights of the "owners of the land and lagoons" in a Lunda polity, see David M. Gordon, *Nachituti's Gift: Economy, Society, and Environment in Central Africa* (Madison: University of Wisconsin Press, 2006), chap. 1.

17. Vansina, "It Never Happened," 387n1.

18. For more details, see Giacomo Macola, *The Kingdom of Kazembe: History and Politics in North-Eastern Zambia and Katanga to 1950* (Hamburg, Germany: LIT Verlag, 2002), chaps. 2–3.

19. Pedro João Baptista, "Journey of the 'Pombeiros' from Angola to the Rios de Senna," in *The Lands of Cazembe*, ed. Richard F. Burton (London: John Murray, 1873), 170.

20. Mutumba Mainga, *Bulozi under the Luyana Kings: Political Evolution and State Formation in Pre-Colonial Zambia* (London: Longman, 1973), 34–35; Eugene L. Hermitte, "An Economic History of Barotseland, 1800–1940" (PhD diss., Northwestern University, 1974), 26.

21. This is one of the central findings of Jack Hogan, "The Ends of Slavery in Barotseland, Western Zambia (c. 1800–1925)" (PhD diss., University of Kent, 2014).

22. Mainga, *Bulozi*, 36.

23. Ibid., 39.

24. Roberts, *History of Zambia*, 97.

25. Joseph C. Miller, *Way of Death: Merchant Capitalism and the Angolan Slave Trade, 1730–1830* (Madison: University of Wisconsin Press, 1988), chap. 2.

26. Vansina's remarks about the Kuba royal capitals, to the immediate north of our study area, are relevant in this context. Jan Vansina, *The Children of Woot: A History of the Kuba Peoples* (Madison: University of Wisconsin Press, 1978), 168–69.

27. Reefe, "Societies," 177–78.

28. Antonio C. P. Gamitto, *King Kazembe and the Marave, Cheva, Bisa, Bemba, Lunda and Other Peoples of Southern Africa*, trans. Ian G. Cunnison, 2 vols. (Lisbon: Junta de Investigações do Ultramar, 1960), 2:15. For an in-depth study of eastern Lunda urbanism, see Giacomo Macola, "The History of the Eastern Lunda Royal Capitals to 1900," in *The Urban Experience in Eastern Africa, c. 1750–2000*, ed. Andrew Burton, special issue, *Azania* 14, nos. 36–37 (2002): 31–45.

29. Reefe, *The Rainbow and the Kings*, 91–92.

30. Gwyn Prins, *The Hidden Hippopotamus: Reappraisal in African History: The Early Colonial Experience in Western Zambia* (Cambridge: Cambridge University Press, 1980), 165.

31. Cf., e.g., Leroy Vail, ed., *The Creation of Tribalism in Southern Africa* (London: James Currey, 1989), and Ronald R. Atkinson, *The Roots of Ethnicity: The Origins of the Acholi of Uganda before 1800* (Philadelphia: University of Pennsylvania Press, 1994).

32. Andrew D. Roberts, *A History of the Bemba: Political Growth and Change in North-Eastern Zambia before 1900* (London: Longman, 1973).

33. Robert J. Papstein, "The Upper Zambezi: A History of the Luvale People, 1000–1900" (PhD diss., UCLA, 1978), chap. 5.

34. François Coillard, *On the Threshold of Central Africa: A Record of Twenty Years' Pioneering among the Barotse of the Upper Zambesi* (1897; repr., London: Frank Cass, 1971), 604.

35. Achim von Oppen, *Terms of Trade and Terms of Trust: The History and Contexts of Pre-Colonial Market Production around the Upper Zambezi and Kasai* (Munich: LIT Verlag, n.d. [1994?]), 32–33; Robert J. Papstein, "From

Ethnic Identity to Tribalism: The Upper Zambezi Region of Zambia, 1830–1981," in *The Creation of Tribalism in Southern Africa*, ed. Leroy Vail (London: James Currey, 1989), 372–94.

36. Cf. Edward A. Alpers, *Ivory and Slaves in East Central Africa: Changing Patterns of International Trade to the Later Nineteenth Century* (London: Heinemann, 1975), 180, and Roberts' more skeptical take in *History of the Bemba*, 193n127.

37. The extent to which coastal activity resulted in the marginalization of independent Nyamwezi traders has been debated by historians. Cf. Abdul Sheriff, *Slaves, Spices and Ivory in Zanzibar: Integration of an East African Commercial Empire into the World Economy, 1770–1873* (London: James Currey, 1987), 172–83, with Stephen J. Rockel, "'A Nation of Porters': The Nyamwezi and the Labour Market in Nineteenth-Century Tanzania," *Journal of African History* 41, no. 2 (2000): 173–95. See also Jan-Georg Deutsch, "Notes on the Rise of Slavery and Social Change in Unyamwezi, c. 1860–1900," in *Slavery in the Great Lakes Region of East Africa*, ed. Henri Médard and Shane Doyle (Oxford: James Currey, 2007), 76–110.

38. Gamitto, *King Kazembe*, 2:25, 87.

39. Said ibn Habib, "Narrative of Said bin Habeeb, an Arab Inhabitant of Zanzibar," *Transactions of the Bombay Geographical Society* 15 (1860): 146. Cf. Roberts, "Firearms in North-Eastern Zambia before 1900," 5.

40. Emrys Chew, *Arming the Periphery: The Arms Trade in the Indian Ocean during the Age of Global Empire* (New York: Palgrave Macmillan, 2012), 41–42. The figures—based on an oft-quoted 1888 report by Euan-Smith, the British consul in Zanzibar—are to be found in Iliffe, *Modern History of Tanganyika*, 51.

41. Habib, "Narrative of Said bin Habeeb," 147.

42. François Bontinck, "La double traversée de l'Afrique par trois 'Arabes' de Zanzibar (1845–1860)," *Études d'Histoire Africaine* 6 (1974): 16; David Livingstone, *Livingstone's Private Journals, 1851–1853*, ed. Isaac Schapera (London: Chatto & Windus, 1960), 43, 228.

43. For all of the above, see Alpers, *Ivory and Slaves*, chaps. 3–7. See also his "The Yao in Malawi: The Importance of Local Research," in *The Early History of Malawi*, ed. Bridglal Pachai (London: Longman, 1972), 171–72.

44. Harry W. Langworthy, "Swahili Influence in the Area between Lake Malawi and the Luangwa River," *African Historical Studies* 4, no. 3 (1971): 583, 585.

45. David Livingstone, *The Last Journals of David Livingstone, in Central Africa, from 1865 to His Death*, ed. Horace Waller, 2 vols. (London: John Murray, 1874), vol. 1, chap. 3.

46. Roberts, "Firearms in North-Eastern Zambia before 1900," 7.

47. Tippu Tip, *Maisha ya Hamed bin Muhammed el Murjebi, yaani Tippu Tip, kwa maneno yake mwenyewe*, ed. and trans. Wilfred H. Whiteley (1958–59; repr., Nairobi: East African Literature Bureau, 1974), 23.

48. Roberts, *History of the Bemba*, 153–60.

49. Marcia Wright and Peter Lary, "Swahili Settlements in Northern Zambia and Malawi," *African Historical Studies* 4, no. 3 (1971): 554.

50. Macola, *Kingdom of Kazembe*, chap. 5.

51. On the history of Benguela, see Mariana P. Candido's recent study, *An African Slaving Port and the Atlantic World: Benguela and Its Hinterland* (Cambridge: Cambridge University Press, 2013).

52. Jan Vansina, "Long-Distance Trade-Routes in Central Africa," *Journal of African History* 3, no. 3 (1962): 383–84; Miller, *Way of Death*, 146, 238–39; Candido, *African Slaving Port*, 171–75, 258–62, 300–302, 304–5.

53. Livingstone, *Private Journals*, 203; David Livingstone, *Missionary Travels and Researches in South Africa* (New York: Harper & Brothers, 1858), 105–6.

54. von Oppen, *Terms of Trade*, 59–86.

55. Linda M. Heywood, "Slavery and Forced Labor in the Changing Political Economy of Central Angola, 1850–1949," in *The End of Slavery in Africa*, ed. Suzanne Miers and Richard Roberts (Madison: University of Wisconsin Press, 1988), 417.

56. Vellut, "Notes sur le Lunda," 94; Miller, *Way of Death*, 145, 214.

57. Miller, *Way of Death*, 214–15.

58. Vellut, "Notes sur le Lunda," 99–110.

59. Ibid., 92–93, 145–46.

60. The standard work on Chokwe trade and expansion remains Joseph C. Miller, "Cokwe Trade and Conquest in the Nineteenth Century," in *Pre-Colonial African Trade: Essays on Trade in Central and Eastern Africa before 1900*, ed. Richard Gray and David Birmingham (London: Oxford University Press, 1970), 175–201.

61. Sebestyen and Vansina, "Angola's Eastern Hinterland in the 1750s," 347.

62. László [Ladislaus] Magyar, "Ladislaus Magyar's erforschung von Inner-Afrika," *Petermann's geographische Mitteilungen* 6 (1860): 231. A Portuguese translation of this important source is available in Isabel de Castro Henriques, *Commerce et changement en Angola au XIXe siècle: Imbangala et Tshokwe face à la modernité*, 2 vols. (Paris: L'Harmattan, 1995), 2:247–61. On Ruund and firearms, see also David Birmingham, *Central Africa to 1870: Zambezi, Zaïre and the South Atlantic* (Cambridge: Cambridge University Press, 1981), 115.

63. Henrique A. Dias de Carvalho, *Expedição portugueza ao Muatiân-vua: Ethnographia e historia tradicional dos povos da Lunda* (Lisbon: Imprensa Nacional, 1890), 656.

64. Miller, "Cokwe Trade," 198.

65. Verney L. Cameron, *Across Africa* (New York: Harper & Brothers, 1877), 341.

66. David M. Gordon, "The Abolition of the Slave Trade and the Transformation of the South-Central African Interior during the Nineteenth Century," *William and Mary Quarterly*, 3rd ser., 66, no. 4 (2009): esp. 925–31.

67. Roberts, "Firearms in North-Eastern Zambia before 1900," 16; Feierman, *Shambaa Kingdom*, 142.

68. Allen F. Roberts, A *Dance of Assassins: Performing Early Colonial Hegemony in the Congo* (Bloomington: Indiana University Press, 2013).

69. On the *ruga ruga* of Nyamwezi warlord Mirambo, see Richard J. Reid, *War in Pre-Colonial Eastern Africa: The Patterns and Meanings of State-Level Conflict in the Nineteenth Century* (Oxford: James Currey, 2007), 99, 144–45, 158–59, and Pesek, "*Ruga-ruga.*"

70. Miller, *Way of Death*, chap. 5.

71. Feierman, *Shambaa Kingdom*, 166, 172–82.

72. See, especially, David Gordon's stimulating but empirically problematic "Abolition of the Slave Trade," and, by the same author, "Wearing Cloth, Wielding Guns: Consumption, Trade, and Politics in the South Central African Interior during the Nineteenth Century," in *The Objects of Life in Central Africa: The History of Consumption and Social Change, 1840–1980*, ed. Robert Ross, Marja Hinfelaar, and Iva Peša (Leiden: Brill, 2013), 17–39.

73. Ian Phimister, *An Economic and Social History of Zimbabwe, 1890–1948: Capital Accumulation and Class Struggle* (London: Longman, 1988), 17.

74. Roberts, *History of the Bemba*, 206.

75. On the origins and workings of the lower Zambezi's *prazos da coroa*, see Allen F. Isaacman, *Mozambique: The Africanization of a European Institution: The Zambesi Prazos, 1750–1902* (Madison: University of Wisconsin Press, 1972), chap. 2; and Malyn D. D. Newitt, *Portuguese Settlement on the Zambesi: Exploration, Land Tenure and Colonial Rule in East Africa* (London: Longman, 1973), chaps. 4, 6–8.

76. Allen F. Isaacman and Barbara S. Isaacman, *Slavery and Beyond: The Making of Men and Chikunda Ethnic Identities in the Unstable World of South-Central Africa, 1750–1920* (Portsmouth, NH: Heinemann, 2004), 39–40.

77. Ibid., 58, 66.

78. Alpers, *Ivory and Slaves*, 216–17.

79. Isaacman and Isaacman, *Slavery and Beyond*, 70–71.

80. Ibid., chaps. 3–7.

81. To my knowledge, the most persuasive take on the complex causes of the Mfecane is still Elizabeth A. Eldredge, "Sources of Conflict in Southern Africa, c. 1800–30: The 'Mfecane' Reconsidered," *Journal of African History* 33, no. 1 (1992): 1–35.

82. John Wright, "Turbulent Times: Political Transformations in the North and East, 1760s–1830s," in *The Cambridge History of South Africa*, vol. 1, *From Early Times to 1885*, ed. Carolyn Hamilton, Bernard K. Mbenga, and Robert Ross (Cambridge: Cambridge University Press, 2010), 221–23.

CHAPTER 2: THE DOMESTICATION OF THE MUSKET ON THE UPPER ZAMBEZI

1. Richard J. Reid, *War in Pre-Colonial Eastern Africa: The Patterns and Meanings of State-Level Conflict in the Nineteenth Century* (Oxford: James Currey, 2007), 11.

2. Humphrey J. Fisher and Virginia Rowland, "Firearms in the Central Sudan," *Journal of African History* 12, no. 2 (1971): 227–28; Gerald M. Berg, "The Sacred Musket: Tactics, Technology, and Power in Eighteenth-Century Madagascar," *Comparative Studies in Society and History* 27, no. 2 (1985): 265–66.

3. Joseph C. Miller, *Way of Death: Merchant Capitalism and the Angolan Slave Trade, 1730–1830* (Madison: University of Wisconsin Press, 1988), 87–88.

4. Achim von Oppen, *Terms of Trade and Terms of Trust: The History and Contexts of Pre-Colonial Market Production around the Upper Zambezi and Kasai* (Munich: LIT Verlag, n.d. [1994?]), 169–73.

5. Miller, *Way of Death*, 93.

6. According to Miller, some sixty thousand muskets per year are likely to have entered the coastal strip running from Loango to Benguela in the latter part of the eighteenth century. Miller, estimating that only one-third of these guns would have been in functioning order at any given time, views such numbers as insufficient to seriously affect socio-military life in the Angolan interior. This chapter, as will become clear, takes a different approach. Besides, Miller is himself somewhat hesitant, admitting, for instance, that "it need not have taken many guns . . . to produce a pervasive atmosphere of danger in limited areas" and that "hunters who were also smithies" did benefit from the introduction of firearms. Ibid., 89, 91–92.

7. Document dated 10 March 1800, cited in Mariana P. Candido, "Merchants and the Business of the Slave Trade at Benguela, 1750–1850," *African Economic History* 35, no. 1 (2007): 16.

8. Maria Emilia Madeira Santos, "Tecnologias em presença: Manufacturas Europeias e artefactos Africanos (c. 1850–1880)," in *Actas de I reunião internacional de história de África: Relação Europa-África no 3° quartel do séc. XIX*, ed. Maria Emilia Madeira Santos (Lisbon: Instituto de Investigação Científica e Tropical, 1989), esp. 221. For similar observations in a different context, see Clapperton C. Mavhunga, "Firearms Diffusion, Exotic and Indigenous Knowledge Systems in the Lowveld Frontier, South Eastern Zimbabwe, 1870–1920," *Comparative Technology Transfer and Society* 1, no. 2 (2003): 212.

9. Colleen E. Kriger, *Pride of Men: Ironworking in 19th Century West Central Africa* (Portsmouth, NH: Heinemann, 1999), 11–12, 58–59.

10. Kriger's otherwise splendid monograph has comparatively little to say about firearms and is certainly mistaken in asserting that gun-mending skills in west-central Africa were restricted to the Chokwe; ibid., 131.

11. Rory Pilossof, "'Guns Don't Colonise People . . .': The Role and Use of Firearms in Pre-Colonial and Colonial Africa," *Kronos* 36, no. 1 (2010): 270. See also John K. Thornton, *Warfare in Atlantic Africa, 1500–1800* (London: UCL Press, 1999), 151.

12. Nelly E. J. Oudshoorn and Trevor J. Pinch, "Introduction: How Users and Non-Users Matter," in *How Users Matter: The Co-Construction of Users and Technology*, ed. Nelly E. J. Oudshoorn and Trevor J. Pinch (Cambridge, MA: MIT Press, 2005), 1–2.

13. Mavhunga, "Firearms Diffusion," 204.

14. Walima T. Kalusa, "Elders, Young Men, and David Livingstone's 'Civilizing Mission': Revisiting the Disintegration of the Kololo Kingdom, 1851–1864," *International Journal of African Historical Studies* 42, no. 1 (2009): 67–68.

15. David Livingstone, *Livingstone's Private Journals, 1851–1853*, ed. Isaac Schapera (London: Chatto & Windus, 1960), 194, 215, 237. See also David Livingstone, *Livingstone's African Journal, 1853–1856*, ed. Isaac Schapera, 2 vols. (London: Chatto & Windus, 1963), 2:294; and Antonio Francisco da Silva Porto, *Viagens e apontamentos de um portuense em África* (Lisbon: Agência Geral das Colónias, 1942), 121, 150.

16. David Livingstone, *Missionary Travels and Researches in South Africa* (New York: Harper & Brothers, 1858), 106, 209.

17. For comments on Kololo marksmanship, see ibid., 228, 279–80. Having repeatedly been asked for "gun medicine," Livingstone eventually volunteered to teach the Kololo paramount, Sekeletu, how to shoot. *Private Journals*, 143, 147.

18. Ibid., 143.

19. Cf. ibid., 232 and Livingstone, *Missionary Travels*, 235–36.

20. James Chapman, *Travels in the Interior of South Africa, 1849–1863*, ed. E. C. Tabler, 2 vols. (1868; repr., Cape Town: A. A. Balkema, 1971), 1:116, 114.

21. Livingstone, *Missionary Travels*, 217.

22. David Livingstone and Charles Livingstone, *Narrative of an Expedition to the Zambezi and Its Tributaries* (London: John Murray, 1865), 292; Andrew D. Roberts, *A History of Zambia* (London: Heinemann, 1976), 127; Mutumba Mainga, *Bulozi under the Luyana Kings: Political Evolution and State Formation in Pre-Colonial Zambia* (London: Longman, 1973), 84.

23. Livingstone, *Private Journals*, 16–17.

24. Livingstone, *African Journal*, 2:331.

25. Ibid., 2:203, and Livingstone, *Missionary Travels*, 105–6.

26. Maria Emilia Madeira Santos, "Introdução (trajectória do comércio do Bié)," in *Viagens e apontamentos de um portuense em África*, ed. Maria Emilia Madeira Santos (Coimbra: Biblioteca Geral da Universidade de Coimbra, 1986), 114–17.

27. Alexandre da Silva Teixeira, "Relação da viagem que fiz deste cidade de Benguella para as terras de Lovar no anno de mil setecentos noventa e quatro," *Arquivos de Angola* (Luanda) 1, no. 4 (1935), doc. X. See also von Oppen, *Terms of Trade*, 176.

28. Alexandra da Silva Teixeira (?), "Derrota de Benguella para o sertão," in *Angola: Apontamentos sôbre a colonização dos planaltos e litoral do sul de Angola*, ed. Alfredo de Albuquerque Felner, 3 vols. (Lisbon: Agência Geral das Colónias, 1940), 2:25. The author of the "Derrota"—which dates to the early 1800s—was almost certainly the already cited Silva Teixeira; François Bontinck, "Derrota de Benguella para o sertão: critique d'authenticité," *Bulletin des Séances de l'Académie Royale des Sciences d'Outre-Mer*, n.s., 23, no. 3 (1977): 279–300.

29. Livingstone, *African Journal*, 2:270; *Private Journals*, 42.

30. von Oppen, *Terms of Trade*, 59–60.

31. A. E. Horton, comp., *A Dictionary of Luvale* (El Monte, CA: Rahn Brothers Printing and Lithographing, 1953).

32. László [Ladislaus] Magyar, "Ladislaus Magyar's erforschung von Inner-Afrika," *Petermann's geographische Mitteilungen* 6 (1860): 233.

33. Mose K. Sangambo, *The History of the Luvale People and Their Chieftainship*, 2nd ed. (Zambezi, Zambia: Mize Palace, n.d. [1985?]), 44; von Oppen, *Terms of Trade*, 367.

34. Horton, *Dictionary of Luvale*, s.v. "-ivwi."

35. Jeff Ramsay, "Firearms in Nineteenth-Century Botswana: The Case of Livingstone's 8-Bore Bullet," *South African Historical Journal* 66, no. 3 (2014): 444.

36. William K. Storey, *Guns, Race, and Power in Colonial South Africa* (Cambridge: Cambridge University Press, 2008), 140.

37. Livingstone, *Private Journals*, 42.

38. Magyar, "Erforschung," 229. The above English translation is to be found in von Oppen, *Terms of Trade*, 173. See also Joseph C. Miller, "Cokwe Trade and Conquest in the Nineteenth Century," in *Pre-Colonial African Trade: Essays on Trade in Central and Eastern Africa before 1900*, ed. Richard Gray and David Birmingham (London: Oxford University Press, 1970), 176, and, for a later period, Charles M. N. White, *The Material Culture of the Lunda-Lovale Peoples*, Occasional Papers of the Rhodes-Livingstone Museum 3 (Livingstone, Zambia: Rhodes-Livingstone Museum, 1948), 5.

39. "Manjeam sofrìvelmente as armas de fogo que, com excepção do cano, fabricam perfeitamente, melhor que nenhuma outra tribu." Silva Porto, *Viagens*, 136–37; Maria Emilia Madeira Santos, "Tecnologias em presença: Manufacturas Europeias e artefactos Africanos (c. 1850–1880)," in *Actas de I reunião internacional de história de África: Relação Europa-África no 3° quartel do séc. XIX*, ed. Maria Emilia Madeira Santos (Lisbon: Instituto de Investigação Científica e Tropical, 1989), 226.

40. Isabel de Castro Henriques, *Commerce et changement en Angola au XIXe siècle: Imbangala et Tshokwe face à la modernité*, 2 vols. (Paris: L'Harmattan, 1995), 1:301, and Miller, "Cokwe Trade and Conquest," 176. According to Headrick, African ironsmiths could not manufacture barrels because the wrought iron they had at their disposal was "not consistent enough." Contrary to what Headrick assumed, however, this did not prevent them from producing "precision parts." Daniel R. Headrick, *The Tools of Empire: Technology and European Imperialism in the Nineteenth Century* (Oxford: Oxford University Press, 1981), 108.

41. Livingstone, *Private Journals*, 42.

42. On the caliber of muskets, see Headrick, *Tools of Empire*, 99, and Paul Dubrunfaut, "Trade Guns in Africa," in *Fatal Beauty: Traditional Weapons from Central Africa* (Taipei: National Museum of History, 2009), 83. On gunflints, see David W. Phillipson, "Gun-Flint Manufacture in North-Western Zambia," *Antiquity* 43, no. 172 (1969): 301–4.

43. Horton, *Dictionary of Luvale.*

44. Ibid., s.v. "-ta."

45. von Oppen, *Terms of Trade,* 172. On the importance of local *aprendizagem* and *treino,* see Santos, "Tecnologias em presença," 221–22.

46. Livingstone, *African Journal,* 1:45.

47. Ibid., 1:43, 48–49.

48. Ibid., 2:270, 271–72.

49. See, e.g., Livingstone, *Missionary Travels,* 211.

50. Livingstone, *Private Journals,* 245.

51. Magyar, "Erforschung," 234; Santos, "Introdução," 83.

52. Storey, *Guns, Race, and Power,* 78.

53. Livingstone, *Private Journals,* 177. In the much the same vein, Chapman called the gun a "potent peacemaker" in the early 1860s; *Travels in the Interior of South Africa,* 2:149.

54. Robert J. Papstein, "The Upper Zambezi: A History of the Luvale People, 1000–1900" (PhD diss., UCLA, 1978), 171–72.

55. Ibid., 191–92.

56. Direct witnesses of the last stages of Luvale expansion include Verney L. Cameron, *Across Africa* (New York: Harper & Brothers, 1877), 367; Frederick S. Arnot, *Garenganze; or, Seven Years' Pioneer Mission Work in Central Africa* (London: James E. Hawkins, 1889), 159, 161, 165, 248; and D. Crawford, in Frederick S. Arnot, *Bihé and Garenganze; or, Four Years' Further Work and Travel in Central Africa* (London: James E. Hawkins, 1893), 45–46, 51.

57. Livingstone, *African Journal,* 1:55, 56.

58. See, e.g., Livingstone, *African Journal,* 1:37, 44; *Missionary Travels,* 295–96, 302, 304.

59. Livingstone, *African Journal,* 1:53, 51; *Missionary Travels,* 312.

60. Livingstone, *African Journal,* 1:45, 2:264; *Missionary Travels,* 305.

61. Cameron, *Across Africa,* 366; see also von Oppen, *Terms of Trade,* 172.

62. François Coillard, *On the Threshold of Central Africa: A Record of Twenty Years' Pioneering among the Barotse of the Upper Zambesi* (1897; repr., London: Frank Cass, 1971), 610–15. The expression "corporeal discourse" comes from Susan Foster, as cited by Allen F. Roberts, *A Dance of Assassins: Performing Early Colonial Hegemony in the Congo* (Bloomington: Indiana University Press, 2013), 79.

63. C. Harding to Secretary to the Administrator (Bulawayo), "Barotseland, Nyakatoro," 27 March 1900, National Archives of Zambia (NAZ), Lusaka, BSAC/NW/HC4/2/1, vol. 6.

64. Ibid; Colin Harding, *In Remotest Barotseland* (London: Hurst and Blackett, 1904), 80. Harding's mention of "caps," of course, implies that, alongside flintlocks, the Luvale also owned some percussion-lock muskets (an example of which is to be found in figure 2.3, second from right).

65. Silva Porto traveled to Bulozi on an almost yearly basis between 1863 and 1869 (Santos, "Introdução," 149). After a first visit to Barotseland in 1871, the Englishman George Westbeech inaugurated a trading station at

Pandamatenga, some sixty miles to the south of the Victoria Falls, and rapidly became the "most influential European" in the area. Gwyn Prins, *The Hidden Hippopotamus: Reappraisal in African History: The Early Colonial Experience in Western Zambia* (Cambridge: Cambridge University Press, 1980), 174.

66. Antonio Francisco da Silva Porto, "Viagens e apontamentos de um portuense em África," vol. 3, 17 December 1865, Biblioteca pública municipal do Porto, Porto, Portugal, Res. ms 1237.

67. "[N]ão havendo método n'este modo de se exercitaram na arte da destruição." Antonio Francisco da Silva Porto, "Viagens e apontamentos de um portuense em África," vol. 4, 8 June 1867, Biblioteca pública municipal do Porto, Porto, Portugal, Res. ms 1239.

68. Emil Holub, *Seven Years in South Africa: Travels, Researches, and Hunting Adventures between the Diamond-Fields and the Zambesi* (1872–79), trans. Ellen E. Frewer, 2 vols. (London: Sampson Low, Marston, Searle & Rivington, 1881), 2:174, 217, 341–42.

69. Ibid., 2:339–40.

70. Silva Porto, "Viagens e apontamentos," vol. 4, 1 June 1867.

71. Holub, *Seven Years*, 2:244–45, 256–57.

72. Emil Holub, *Emil Holub's Travels North of the Zambezi, 1885–6*, ed. Ladislav Holy, trans. Christa Johns (Manchester: Manchester University Press, 1975), 272–73.

73. George Westbeech, "The Diary of George Westbeech," in *Trade and Travel in Early Barotseland*, ed. E. C. Tabler (London: Chatto & Windus, 1963), 92.

74. Holub, *Seven Years*, 2:134; Arnot, *Garenganze*, 90.

75. See, e.g., Holub, *Seven Years*, 2:125.

76. Ibid., 2:146–47, 142. This is confirmed by Silva Porto, "Viagens e apontamentos," vol. 4, 22 June 1867, and "Viagens e apontamentos de um portuense em África," vol. 5, 2 August 1868, Biblioteca pública municipal do Porto, Porto, Portugal, Res. ms 1240.

77. Holub, *Seven Years*, 2:200.

78. Ibid., 2:142, 160.

79. Ibid., 2:241–42.

80. Ibid., 2:143.

81. Ibid., 2:228, 238.

82. Antonio Francisco da Silva Porto, 29 March 1884, quoted in Santos, "Tecnologias em presença," 224.

83. Holub, *Seven Years*, 2:341–42.

84. Arnot, *Garenganze*, 78.

85. Alexandre A. da Rocha de Serpa Pinto, *How I Crossed Africa*, trans. Alfred Elwes, 2 vols. (London: Sampson Low, Marston, Searle, & Rivington, 1881), 2:9, 22. The same Serpa Pinto, however, is also on record as having stated that, besides obtaining Belgian-made flintlocks from Benguela, the Lozi ("Luinas") also had "a good many percussion-muskets of English manufacture, conveyed thither by the traders from the South." Ibid., 2:40.

86. Ibid., 2:46–47. Doubts about the veracity of Serpa Pinto's account were expressed by Westbeech, "Diary of George Westbeech," 49.

87. Holub, *Emil Holub's Travels*, 205. For the date of Lewanika's first raid against the Ila, see G. Westbeech to F. S. Arnot, Lealui, 5 October 1882, in Arnot, *Garenganze*, 62.

88. P. Berghegge to A. Weld, Pandamatenga, 1 November 1883, in Michael Gelfand, ed., *Gubulawayo and Beyond: Letters and Journals of the Early Jesuit Missionaries to Zambesia (1879–1887)* (London: Geoffrey Chapman, 1968), 419.

89. Arnot, *Garenganze*, 92. But cf. Santos, "Tecnologias em presença," 226.

90. Coillard, *Threshold*, 199; Westbeech, "Diary of George Westbeech," 47; Mainga, *Bulozi*, 127.

91. Coillard, *Threshold*, 261.

92. Holub, *Emil Holub's Travels*, 9. See also Serpa Pinto, *How I Crossed Africa*, 2:41.

93. Holub, *Emil Holub's Travels*, 279.

94. Coillard, *Threshold*, 300.

95. Frederick Courteney Selous, *Travel and Adventure in South-East Africa* (London: Rowland Ward, 1893), 252.

96. Coillard, *Threshold*, 356, 387; Mainga, *Bulozi*, 176; and Prins, *Hidden Hippopotamus*, 220.

97. Gerald L. Caplan, *The Elites of Barotseland, 1878–1969: A Political History of Zambia's Western Province* (London: C. Hurst, 1970), 52, 54.

98. "[É]tonnés de la quantité de fusils qu'il y avait dans le pays, c'est par centaines qu'ils se comptaient. C'étaient surtout des fusils à pierre et des pseudo-mousquets introduits par les ma-Mbari." Adolphe D. Jalla, *Pionniers parmi les ma-Rotse* (Florence: Imprimerie Claudienne, 1903), 100.

99. Papstein, "Upper Zambezi," 195–96.

100. See, e.g., von Oppen, *Terms of Trade*, 244, 354.

101. Adolphe D. Jalla, comp., *English-Sikololo Dictionary* (Torre Pellice, Italy: Imprimerie Alpine, 1917); J. Jeffrey Hoover, email to author, 4 May 2013.

102. Grant D. McCracken, *Culture and Consumption: New Approaches to the Symbolic Character of Consumer Goods and Activities* (Bloomington: Indiana University Press, 1988), 11.

103. Silva Porto, "Viagens e apontamentos," vol. 4, 27 June 1867; Serpa Pinto, *How I Crossed Africa*, 2:4.

104. McCracken, *Culture and Consumption*, 11.

105. Livingstone, *Private Journals*, 43; Livingstone, *African Journal*, 1:11–12; Edwin W. Smith and Andrew M. Dale, *The Ila-Speaking Peoples of Northern Rhodesia*, 2 vols. (London: Macmillan, 1920), 1:33–34, 39; and François Bontinck, "La double traversée de l'Afrique par trois 'Arabes' de Zanzibar (1845–1860)," *Études d'Histoire Africaine* 6 (1974): 15.

106. Frank H. Melland, *In Witch-Bound Africa: An Account of the Primitive Kaonde Tribe and Their Beliefs* (1923; repr., London: Frank Cass, 1967), 273–77.

107. Ibid., 44, 273, 274, 275; Hugues Legros, *Chasseurs d'ivoire: Une histoire du royaume yeke du Shaba (Zaïre)* (Brussels: Éditions de l'Université de Bruxelles, 1996), 92; Dirk Jaeger, email to author, 16 September 2013.

108. Simon Jilundu Chibanza, "Formation of the Kasempa Chieftainship," in Simon Jilundu Chibanza, *Kaonde History,* Part 2 of J. T. Munday and Simon Jilundu Chibanza, *Central Bantu Historical Texts,* vol.1 (Lusaka: Rhodes-Livingstone Institute, 1961), 52, 49.

109. Ibid., 56, 58, 59, 62; Wim van Binsbergen, *Tears of Rain: Ethnicity and History in Central Western Zambia* (London: Kegan Paul International, 1992), 155; E. Copeman to Secretary for Native Affairs, Kasempa, 15 January 1906, NAZ, KDE 2/36/1.

110. Chibanza, "Formation," 63, 59.

111. R. E. Broughall Woods, cited in Melland, *In Witch-Bound Africa,* 272.

112. R. E. Broughall Woods, comp., *A Short Introductory Dictionary of the Kaonde Language with English-Kaonde Appendix* (London: Religious Tract Society, 1924).

113. P. Hall to Secretary for Native Affairs, Kasempa, 23 January 1923, NAZ, BSAC/NR/B1/2/368.

114. C. S. Parsons to P. Hall, Solwezi, 23 January 1923, NAZ, BSAC/NR/B1/2/368.

115. Woods, *Short Introductory Dictionary.*

116. Hall to Secretary for Native Affairs, 23 January 1923.

117. H. G. Pirouet, "'The Gates of Hell Shall Not Prevail,'" *South Africa General Mission Pioneer* 36, no. 4 (April 1923): 42.

CHAPTER 3: THE WARLORD'S MUSKETS

1. Andrew D. Roberts, *A History of Zambia* (London: Heinemann, 1976), 122. The timing of Msiri's establishment in southern Katanga is discussed by Hugues Legros, *Chasseurs d'ivoire: Une histoire du royaume yeke du Shaba (Zaïre)* (Brussels: Éditions de l'Université de Bruxelles, 1996), 28–29. It is regrettable that Dr. Legros (a former general secretary of the Belgian Commission Universitaire pour le Développement) never saw fit to make available to Congolese and other scholars the transcripts of the numerous interviews he conducted in Bunkeya and surrounding areas in the early 1990s.

2. "'Je suis heureux quand je te vois, mon enfant; vois le bien que tu as fait ici. Ce pays était autrefois troublé. Maintenant le pays est à toi! Donne-moi la main!' Ils se serrèrent la main. Mpanda [*sic*] traça une ligne par terre entre eux deux et lui dit: 'Toi, fais aussi une ligne! . . . Je te donne ce pays, à toi-même mon enfant. Je t'aime!'" Mukanda Bantu, "Les mémoires de Mukande Bantu," *Bulletin de la Société Belge d'Études Coloniales* 27, nos. 5–6 (1919): 256–59. See also Legros, *Chasseurs d'ivoire,* 14–17.

3. "Mémoires de Mukande Bantu," 260; Giacomo Macola, *The Kingdom of Kazembe: History and Politics in North-Eastern Zambia and Katanga to 1950* (Hamburg, Germany: LIT Verlag, 2002), 139–40; Legros, *Chasseurs d'ivoire,* 17–18. The approximate date of Lubabila's death can be deduced from David

Livingstone, *The Last Journals of David Livingstone, in Central Africa, from 1865 to His Death*, ed. Horace Waller, 2 vols. (London: John Murray, 1874), 1:276, 297.

4. Edgard Verdick, *Les premiers jours au Katanga (1890–1903)* (Brussels: Comité spécial du Katanga, 1952), 36. This posthumous text (Verdick died in 1927) is based on the officer's original *carnets de route*.

5. For a different chronology, see Legros, *Chasseurs d'ivoire*, 15. Kilolo Ntambo hailed from Kibanda, a Luba polity on the southern borders of the Upemba Depression.

6. Frederick S. Arnot, *Garenganze; or, Seven Years' Pioneer Mission Work in Central Africa* (London: James E. Hawkins, 1889), 231.

7. Joseph A. Moloney, *With Captain Stairs to Katanga* (1893; repr., London: Jeppestown Press, 2007), 125.

8. Mwata Kazembe XIV [Shadreck Chinyanta Nankula], *Central Bantu Historical Texts*, vol. 2, *Historical Traditions of the Eastern Lunda*, trans. Ian G. Cunnison (Lusaka: Rhodes-Livingstone Institute, 1962), 82.

9. Verney L. Cameron, *Across Africa* (New York: Harper & Brothers, 1877), 353.

10. See, for instance, author's interviews with J. Mwidye Kidyamba, Bunkeya, 5 August 2011, and B. Mwenda Numbi, Bunkeya, 7 August 2011.

11. *Gobori*, in turn, is likely to derive from another Swahili word, *koroboi*, a corruption of "carbine." The word *magoba* is to be found in Antoine Mwenda Munongo, ed. and trans., "Chants historiques des Bayeke: Recueillis à Bunkeya et ailleurs," *Problèmes Sociaux Congolais* 77 (1967): 68. My thanks to Drs. Ray Abrahams and Martin Walsh for helping me plot the history of *magoba*.

12. "Car prendre une jeune fille pour sa maison, c'est accroître (sa population comme) des sauterelles." Mwenda Munongo, "Chants historiques des Bayeke," 59–60.

13. Richard J. Reid, *War in Pre-Colonial Eastern Africa: The Patterns and Meanings of State-Level Conflict in the Nineteenth Century* (Oxford: James Currey, 2007), 50.

14. Paul Reichard, "Le Katanga," *Congo* (Brussels) 1, no. 3 (1930): 474.

15. Legros, *Chasseurs d'ivoire*, 114–15.

16. Livingstone, *Last Journals*, 1:308, 321.

17. Ibid., 1:273, 282; François Bontinck, "La double traversée de l'Afrique par trois 'Arabes' de Zanzibar (1845–1860)," *Études d'Histoire Africaine* 6 (1974): 45.

18. Livingstone, *Last Journals*, 1:287, 294; Andrew D. Roberts, *A History of the Bemba: Political Growth and Change in North-Eastern Zambia before 1900* (London: Longman, 1973), 194.

19. Livingstone, *Last Journals*, 1:276.

20. That was certainly the case in 1884; Hermenegildo Carlos de Brito Capello and Roberto Ivens, *De Angola á contra-costa: Descripção de uma viagem atravez do continente africano*, 2 vols. (Lisbon: Imprensa Nacional, 1886), 2:96.

21. Legros, *Chasseurs d'ivoire*, 116, 119. Legros's sources on this point are far from clear and would seem to be partly contradicted by Arnot, *Garenganze*, 233.

22. Capello and Ivens, *De Angola á contra-costa*, 2:106; Arnot, *Garenganze*, 174, 187, 199, 205; D. Crawford, in Frederick S. Arnot, *Bihé and Garenganze; or, Four Years' Further Work and Travel in Central Africa* (London: James E. Hawkins, 1893), 59; Alexandre Delcommune, *Vingt années de vie africaine: Récits de voyages, d'aventures et d'exploration au Congo belge, 1874–1893*, 2 vols. (Brussels: Ferdinand Larcier, 1922), 2:253; and Paul Le Marinel, *Carnets de route dans l'État indépendant du Congo de 1887 à 1910* (Brussels: Éditions Progress, 1991), 194.

23. Arnot, *Garenganze*, 233. See also Legros, *Chasseurs d'ivoire*, 119; Thomas Q. Reefe, *The Rainbow and the Kings: A History of the Luba Empire to 1891* (Berkeley: University of California Press, 1981), 185; and Maria Emilia Madeira Santos, "Introdução (trajectória do comércio do Bié)," in *Viagens e apontamentos de um portuense em África*, ed. Maria Emilia Madeira Santos (Coimbra: Biblioteca Geral da Universidade de Coimbra, 1986).

24. Mukanda Bantu, "Mémoires de Mukande Bantu," 269–70.

25. Arnot, *Garenganze*, 170, 184, 207, 213; Legros, *Chasseurs d'ivoire*, 121.

26. Arnot, *Garenganze*, 174–75, 215.

27. Legros, *Chasseurs d'ivoire*, 102.

28. One of the principal Lualaba crossing points was located near the N'Zilo rapids, close to present-day Kolwezi and then home to "southern Luba" peoples subjugated by the Yeke in the early 1870s. Le Marinel, *Carnets de route*, 155–56, 159. Compare with the itineraries followed by Arnot in 1886 (*Garenganze*, 167–68) and Crawford in 1890 (in Arnot, *Bihé and Garenganze*, 56–57).

29. Cameron, *Across Africa*, 353.

30. Paul Reichard, "Herr Paul Reichard: Bericht über seine Reisen in Ostafrika und dem Quellgebiet des Kongo," *Verhandlungen der Gesellschaft für Erdkunde zu Berlin* 13, no. 2 (1886): 119. See also Paul Reichard, "Bericht von Paul Reichard über die Reise nach Urua und Katanga," *Mittheilungen der Afrikanischen Gesellschaft in Deutschland* 4, no. 5 (1885): 307.

31. Victor Giraud, *Les lacs de l'Afrique équatoriale* (Paris: Librairie Hachette, 1890), 318–19. Giraud was told of Msiri's military might while among the Ushi of the upper Luapula in 1883.

32. "At Msidi's capital"—wrote a clearly fascinated Arnot—"I have met with native traders from Uganda; the Unyamwesi country; the Ungala, to the east of Lake Tanganyika; the Luba country, almost as far down as the Stanley Falls; the basin of the Zambesi; Zumbu, Bihé, and Angola, as well as Arab traders from Lake Nyassa and Zanzibar. Copper, salt, ivory, and slaves are the chief articles of commerce. In exchange for these Msidi purchases flint-lock guns, powder, cloth, and beads, besides many other curious things that these native and Arab traders bring." Arnot, *Garenganze*, 235.

33. "Infeliz d'aquelle que tal intente, porquanto o soba confiscar-lhe-ha tudo, não indemnisando os commerciantes." Capello and Ivens, *De Angola á contra-costa*, 2:104.

34. Moloney, *With Captain Stairs to Katanga*, 127. See also Arnot, *Garenganze*, 235.

35. Legros, *Chasseurs d'ivoire*, chap. 9.

36. Ibid., 80–85.

37. Ibid., 110–11.

38. D. Crawford, "Dec. 1891" [c. 30 December 1891], in the missionary journal *Echoes of Service* 273 (November 1892): 257. This important letter was also published in Arnot, *Bihé and Garenganze*, 116–27.

39. C. A. Swan, 14 December 1889, *Echoes of Service* 230 (January 1891): 25.

40. Interview with B. Mwenda Numbi.

41. C. A. Swan, 9 March 1890, *Echoes of Service* 231 (February 1891): 41.

42. Delcommune, *Vingt années*, 2:250–51.

43. Capello and Ivens, *De Angola á contra-costa*, 2:46, 52.

44. "[I]nsatiable avidité du roi qui, non content de prendre l'ivoire, exigeait de N'Tenke des esclaves, hommes, femmes ou enfants, qu'il vendait aux Bienos ou aux Makangombi." Delcommune, *Vingt années*, 2:329. In this context, "Makangombi" is more likely to designate an important Bihean chief (see, e.g., David Livingstone, *Missionary Travels and Researches in South Africa* [New York: Harper & Brothers, 1858], 238) than a Luvale leader of the same name (see Arnot, *Garenganze*, 155). By the end of the century, the expressions *Kangombés* or *Tungombés* were currently used in southern Congo to refer to Ovimbundu traders. See, for instance, C. Brasseur to D. Brasseur, Lofoi, 20 October 1894 [*sic*, but between 23 and 31 October 1894], Musée Royal de l'Afrique Centrale (MRAC), Tervuren, Belgium, RG 768/81.15, Clément Brasseur's Papers (BP).

45. Legros, *Chasseurs d'ivoire*, 112.

46. C. Brasseur to D. Brasseur, Lofoi, 3 January 1896, MRAC, BP.

47. Capello and Ivens, *De Angola á contra-costa*, 2:121.

48. Cameron, *Across Africa*, 353.

49. Arnot, *Garenganze*, 243.

50. Ibid., 194.

51. D. Crawford, "Dec. 1891," *Echoes of Service* 273 (November 1892): 258.

52. Arnot, *Garenganze*, 242.

53. Legros, *Chasseurs d'ivoire*, 73–74.

54. Capello and Ivens, *De Angola á contra-costa*, 2:169, 173.

55. C. A. Swan, 29 June 1890, *Echoes of Service* 239 (June 1891): 138. On the diffusion of breech-loaders among the Batswana during the last decades of the nineteenth century, see, for instance, Jeff Ramsay, "Firearms in Nineteenth-Century Botswana: The Case of Livingstone's 8-Bore Bullet," *South African Historical Journal* 66, no. 3 (2014): esp. 443.

56. C. A. Swan, 10 April 1890, *Echoes of Service* 231 (February 1891): 42. See also Le Marinel's remarks about Kimpoto and Mirambo's gunmen, above.

57. Interview with B. Mwenda Numbi.

58. Arnot, *Garenganze*, 220–21.

59. "[L]a guerre est devenue dure. (Mais qu'importe, nous irons la faire quand même). Car nous serons bien habillés par notre roi. Par notre suzerain." Mwenda Munongo, "Chants historiques des Bayeke," 63.

60. Arnot, *Garenganze*, 233.

61. Legros, *Chasseurs d'ivoire*, 129–30.

62. For further details, see Macola, *Kingdom of Kazembe*, 151–53.

63. Giraud, *Lacs de l'Afrique équatoriale*, 352, 366.

64. Alfred Sharpe, "Alfred Sharpe's Travels in the Northern Province and Katanga," *Northern Rhodesia Journal* 3, no. 3 (1957): 211. (Published version of Sharpe to Johnston, Mandala, 4 March 1891, encl. in Johnston to Foreign Office, 6 May 1891, NAUK, FO 84/2114.)

65. B. Watson, 30 June 1899, cited in Andrew D. Roberts, "Firearms in North-Eastern Zambia before 1900," *Transafrican Journal of History* 1, no. 2 (1971): 12, 20n89.

66. Reefe, *The Rainbow and the Kings*, 178–80; Legros, *Chasseurs d'ivoire*, 93–94. The most common weapons among the people of the Upemba Depression in the mid-1880s were "light spears, which however are mostly worn as adornments, shields and bows with poisoned arrows" (leichten Speeren, welche jedoch eigentlich mehr zum Schmucke getragen werden, Schild und Bogen mit vergifteten Pfeilen); Reichard, "Bericht über seine reisen in Ostafrika," 117.

67. Reichard, "Bericht über seine reisen in Ostafrika," 121; Reichard, "Bericht über die reise nach Urua und Katanga," 304; and Capello and Ivens, *De Angola á contra-costa*, 2:80.

68. For an admittedly embellished description of a canoe battle in which Luba guile prevailed over Yeke firepower, see Capello and Ivens, *De Angola á contra-costa*, 2:89–90.

69. Reichard, "Bericht über seine reisen in Ostafrika," 115; Arnot, *Garenganze*, 198n; and Le Marinel, *Carnets de route*, 168, 169, 171.

70. Mwenda Munongo, "Chants historiques des Bayeke," 56. The song was entitled "Kwarkasa Kwasa," in imitation of the noise made by the flint when striking the musket's frizzen. The word for flintlock reported by Mwenda Munongo is not *magoba* (see above) but the (perhaps related?) *mabongoya*.

71. Ibid., 135–36.

72. Arnot, *Garenganze*, 239. This may not have been entirely accurate, since, on 8 May 1891, Le Marinel was approached by one of Msiri's men "with a gun to repair" (avec un fusil à réparer). Le Marinel, *Carnets de route*, 193. On Yeke copper bullets, see Eugenia W. Herbert, *Red Gold of Africa: Copper in Precolonial History and Culture* (Madison: University of Wisconsin Press, 1984), 191.

73. C. Brasseur to D. Brasseur, Lofoi, "1 April 1895" [*sic*, but 26 April 1895], MRAC, BP; interview with B. Mwenda Numbi.

74. "[V]ieilles platines d'armes à pierre, vieilles limes, vis sans filet de dimensions diverses, le tout datant pour le moins de l'invention des armes à feu. Je lui vis un jour faire en entier un bois de fusil qui, ma foi, s'adaptait très bien au canon qui lui était destiné." Giraud, *Lacs de l'Afrique équatoriale*, 346.

75. D. Crawford, 23 January 1896, *Echoes of Service* 364 (August 1896): 252.

76. Allen F. Isaacman and Barbara S. Isaacman, *Slavery and Beyond: The Making of Men and Chikunda Ethnic Identities in the Unstable World of South-Central Africa, 1750–1920* (Portsmouth, NH: Heinemann, 2004), 98; T. I. Matthews, "Portuguese, Chikunda, and Peoples of the Gwembe Valley: The Impact of the 'Lower Zambezi Complex' on Southern Zambia," *Journal of African History* 22, no. 1 (1981): 35.

77. A. H. Ackermann to Acting Commercial Representative (BSAC, Bulawayo), n.p., 28 June 1917; and J. W. Sharratt Horne to DC, Chilanga, 4 May 1916, both in NAZ, BSAC/NR/B1/2/262.

78. R. E. Broughall Woods, comp., *A Short Introductory Dictionary of the Kaonde Language with English-Kaonde Appendix* (London: The Religious Tract Society, 1924).

79. David Birmingham, email to author, 26 March 2011.

80. Capello and Ivens, *De Angola á contra-costa*, 2:79.

81. C. A. Swan, 5 June 1889, *Echoes of Service* 222 (June 1890): 184; 18 October 1889, *Echoes of Service* 230 (January 1891): 21; 6 and 28 January 1890, 12 and 26 March 1890, and 29 April 1890, *Echoes of Service* 231 (February 1891): 39, 40, 41, 43; Legros, *Chasseurs d'ivoire*, 121.

82. C. A. Swan, 5 June 1889, *Echoes of Service* 222 (June 1890); and C. A. Swan, 1 November 1899, *Echoes of Service* 230 (January 1891): 21.

83. For detailed treatments of the Yeke wars against Shimba, see Macola, *Kingdom of Kazembe*, 155–60, and Mukanda Bantu, "Mémoires de Mukande Bantu," 271–72.

84. See, e.g., C. A. Swan, 10, 21, 26, and 28 February, and 1 March 1890, *Echoes of Service* 251 (December 1891): 288–89; and Mukanda Bantu "Mémoires de Mukande Bantu," 273–76.

85. D. Crawford, "Dec. 1891," *Echoes of Service* 273 (November 1892): 257. See also D. Crawford, 18 May 1891, *Echoes of Service* 252 (December 1891): 299.

86. Delcommune, *Vingt années*, 2:278–79.

87. K. Asani bin Katompa, "L'opposition Sanga à Msiri et à l'administration coloniale Belge (1891–1911)" (mémoire de licence en histoire, Université Nationale du Zaïre, Lubumbashi, 1977), 37, 42–43.

88. D. Crawford, 18 May 1891, *Echoes of Service* 252 (December 1891): 299.

89. C. A. Swan, 21 February 1891, *Echoes of Service* 251 (December 1891): 288; Legros, *Chasseurs d'ivoire*, 84.

90. C. A. Swan, 26 February 1891, *Echoes of Service* 251 (December 1891): 289.

91. C. A. Swan, 20 March 1891, *Echoes of Service* 251 (December 1891): 290; C. A. Swan, 22 April 1891, *Echoes of Service* 253 (January 1892): 9.

92. C. A. Swan, 20 March 1891, *Echoes of Service* 251 (December 1891): 290.

93. C. A. Swan, 30 March 1891, *Echoes of Service* 251 (December 1891): 291; Le Marinel, *Carnets de route*, 201.

94. C. A. Swan, 10 May 1891, *Echoes of Service* 253 (January 1892): 10; Le Marinel, *Carnets de route*, 194. The mpande—a conus-shell disk—was a Yeke symbol.

95. Le Marinel, *Carnets de route*, 202.

96. D. Crawford, "Dec. 1891," *Echoes of Service* 273 (November 1892): 257. In March, Msiri failed to extort some powder from the Plymouth Brethren missionaries on at least two occasions; C. A. Swan, 11 and 31 March 1891, *Echoes of Service* 251 (December 1891): 289, 291.

97. Le Marinel, *Carnets de route*, 190, 191; C. A. Swan, 4 May 1891, *Echoes of Service* 253 (January 1892): 9.

98. D. Crawford, "Dec. 1891," *Echoes of Service* 273 (November 1892): 257; Mukanda Bantu, "Mémoires de Mukande Bantu," 277.

99. D. Crawford, "Dec. 1891," *Echoes of Service* 273 (November 1892): 257; H. B. Thompson, 8 December 1891, in Arnot, *Bihé and Garenganze*, 112.

100. Delcommune, *Vingt années*, 2:249. See also Le Marinel, *Carnets de route*, 204.

101. A. Legat to A. Delcommune, Lofoi, 13 October 1891, in Delcommune, *Vingt années*, 2:272; D. Crawford, "Dec. 1891," *Echoes of Service* 273 (November 1892): 257.

102. Paul Briart, *Aux sources du fleuve Congo: Carnets du Katanga (1890–1893)*, ed. Dominique Ryelandt (Paris: L'Harmattan, 2003), 177, 179. But cf. Delcommune, *Vingt années*, 2:265.

103. Delcommune, *Vingt années*, 2:274, 295, 298.

104. Ibid., 2:296–97; William G. Stairs, "De Zanzibar au Katanga: Journal du Capitaine Stairs (1890–1891)," *Le Congo Illustré* 2, no. 23 (1893): 183.

105. Delcommune, *Vingt années*, 2:260, 276, 291; Legat to Delcommune, 13 October 1891, ibid., 271–73; Stairs, "De Zanzibar au Katanga," *Le Congo Illustré* 2, no. 23 (1893): 183.

106. William G. Stairs, "De Zanzibar au Katanga: Journal du Capitaine Stairs (1890–1891)," *Le Congo Illustré* 2, no. 25 (1893): 197.

107. H. B. Thompson, 12–27 November 1891, in Arnot, *Bihé and Garenganze*, 110–11.

108. H. B. Thompson, 2 December 1891, ibid., 111–12; Mukanda Bantu, "Mémoires de Mukande Bantu," 498.

109. William G. Stairs, "De Zanzibar au Katanga: Journal du Capitaine Stairs (1890–1891)," *Le Congo Illustré* 2, no. 24 (1893): 191; "De Zanzibar au Katanga," *Le Congo Illustré* 2, no. 25 (1893): 198; Moloney, *With Captain Stairs to Katanga*, 132; D. Crawford, "Dec. 1891," *Echoes of Service* 273 (November 1892): 260, 261.

110. Stairs, "De Zanzibar au Katanga," *Le Congo Illustré* 2, no. 25 (1893): 199. Bodson himself was wounded by Msiri's gunmen and died shortly thereafter.

111. Stairs, "De Zanzibar au Katanga," *Le Congo Illustré* 2, no. 24 (1893): 191. Local interviewees showed no knowledge of this particular offspring of Msiri.

112. William G. Stairs, "De Zanzibar au Katanga: Journal du Capitaine Stairs (1890–1891)," *Le Congo Illustré* 2, no. 26 (1893): 205.

113. "La haine qu'ils ont contre ces détrousseurs est des plus vivace." Delcommune, *Vingt années*, 2:315.

114. Mwenda Munongo, "Chants historiques des Bayeke," 64.

115. D. Crawford, 25 June 1892, *Echoes of Service* 281 (March 1893): 56–57.

CHAPTER 4: GUN SOCIETIES UNDONE?

1. Hugues Legros, *Chasseurs d'ivoire: Une histoire du royaume yeke du Shaba (Zaïre)* (Brussels: Éditions de l'Université de Bruxelles, 1996), 138–41.

2. North-Western Rhodesia (or, to use the diction of the 1899 Order in Council, "Barotziland — North-Western Rhodesia") formed a distinctive administrative unit between the late 1890s and 1911, the year in which it was amalgamated with another BSAC-run territory, North-Eastern Rhodesia. Until 1905, when its border was shifted eastwards, North-Western Rhodesia comprised the territory roughly corresponding to present-day Zambia's Western (Barotseland), North-Western, and Southern Provinces.

3. For a still useful introduction to the origins and workings of the Congo Free State's *régime domanial*, see Jean Stengers and Jan Vansina, "King Leopold's Congo, 1886–1908," in *The Cambridge History of Africa*, vol. 6, *c. 1870–c. 1905*, ed. Roland Oliver and G. N. Sanderson (Cambridge: Cambridge University Press, 1985), 315–58. See also Jan Vansina, *Being Colonized: The Kuba Experience in Rural Congo, 1880–1960* (Madison: University of Wisconsin Press, 2010), esp. 59–60.

4. See Proclamation no. 18 of 1901, in *Cape of Good Hope Government Gazette* 8,370 (10 September 1901).

5. For a recent revisitation of Lozi diplomacy during the Scramble, see Jack Hogan, "'What Then Happened to Our Eden?': The Long History of Lozi Secessionism, 1890–2013," *Journal of Southern African Studies* 40, no. 5 (2014): 907–24.

6. P. J. Macdonnel to Secretary to the Administrator (Northern Rhodesia) (henceforth S. Admin.), Livingstone, 20 June 1912, NAZ, BSAC/NR/B1/2/368.

7. DC (Batoka), "A System for the Voluntary Registration of Native Guns," encl. in DC (Batoka) to R. T. Coryndon, Kalomo, 22 June 1903, NAZ, BSAC/NW/A3/30.

8. Government Notice no. 4 of 1901, in *British Central Africa Gazette* 8, no. 1 (31 January 1901). See also Stuart A. Marks, *Large Mammals and a Brave People: Subsistence Hunters in Zambia* (Seattle: University of Washington Press, 1976), 73.

9. Government Notice no. 4 of 1900, in *British Central Africa Gazette* 7, no. 12 (31 December 1900); Government Notice no. 9 of 1902, in *British Central Africa Gazette* 9, no. 8 (31 August 1902).

10. Proclamation no. 1 of 1905, in *North-Eastern Rhodesia Government Gazette* 3, no. 4 (29 April 1905).

11. Dugald Campbell, *In the Heart of Bantuland* (London: Seeley, Service, 1922), 28–29; Jean-Luc Vellut, "Garenganze/Katanga–Bié–Benguela and Beyond: The Cycle of Rubber and Slaves at the Turn of the 20th Century," *Portuguese Studies Review* 19, nos. 1–2 (2011): 139, 146–48.

12. See, e.g., René Pélissier, *Les guerres grises: Résistance et révoltes en Angola, 1845–1941* (Orgeval, France: Pélissier, 1977), chap. 16.

13. Vice-Governor General to "Secrétaire d'État," Boma, 26 April 1907, Archives africaines de l'ex-ministère des Affaires africaines (AA), Ministère des Affaires étrangères, Brussels, Belgium, AE (301) 405. For suspicions of Portuguese officials, see, for instance, A. St. H. Gibbons to Director of Military Intelligence, "Zambesi River," 2 October 1899, and C. Harding to Secretary (BSAC), Lealui, 4 July 1900, both in NAZ, KDE 2/44/1–3; and Director of Military Intelligence to Colonial Office, [London], 24 April 1901, encl. in J. Chamberlain to High Commissioner for South Africa (henceforth HCSA), London, 31 May 1901, NAZ, BSAC/NW/HC 1/2/1.

14. See, e.g., DC (Lealui) to S. Admin., Lealui, 4 February 1904, and Acting S. Admin. to Acting DC (Lealui), Kalomo, 7 March 1904, both in NAZ, BSAC/NW/A3/24/9. See also Lewis H. Gann, "The End of the Slave Trade in British Central Africa, 1889–1912," *Rhodes-Livingstone Journal* 16 (1954): 49.

15. C. Harding to Secretary (BSAC), Lealui, 25 June 1900, NAZ, BSAC/NW/A6/1/1.

16. C. Harding to Secretary (BSAC), Lealui, 4 July 1900, NAZ, KDE 2/44/1–3.

17. Gerald L. Caplan, *The Elites of Barotseland, 1878–1969: A Political History of Zambia's Western Province* (London: C. Hurst, 1970), chap. 4.

18. F. Aitkens to R. T. Coryndon, Lealui, 17 March 1903, and S. Admin. to F. Aitkens, Kalomo, 14 August 1903, both in NAZ, BSAC/NW/A3/3/2.

19. F. Aitkens to S. Admin., Lealui, 7 March 1906, NAZ, BSAC/NW/A3/3/2.

20. See, e.g., DC (Falls District) to S. Admin., Livingstone, 4 August 1903, NAZ, BSAC/NW/A3/30; F. Aitkens to Secretary for Native Affairs (henceforth SNA), Lealui, 24 August 1905, NAZ, BSAC/NW/A3/24/9.

21. L. A. Wallace to H. J. Gladstone, Livingstone, 7 February 1912, NAZ, BSA/NR/HC1/3/1, vol. 2.

22. D. Thwaits to C. McKinnon, Njamba's, 27 November 1911, encl. in L. A. Wallace to H. J. Gladstone, Livingstone, 7 February 1912, NAZ, BSA/NR/HC1/3/1, vol. 2.

23. A. St. H. Gibbons to Director of Military Intelligence, "Zambesi River," 2 October 1899, and C. Harding to Secretary (BSAC), Lealui, 4 July 1900, both in NAZ, KDE 2/44/1–3.

24. C. Harding to Secretary to the Administrator (Bulawayo), "Barotseland, Nyakatoro," 27 March 1900, NAZ, BSAC/NW/HC4/2/1, vol. 6.

25. J. H. Venning, "Early Days in Balovale," *Northern Rhodesia Journal* 2, no. 6 (1955): 55.

26. Ibid., 57.

27. R. H. Palmer to C. McKinnon, Balovale, 31 August 1911, encl. in C. McKinnon to L. A. Wallace, Sesheke, 3 October 1911, NAZ, BSAC/NR/HC1/3/1, vol. 1.

28. Alfred St. H. Gibbons, *Africa from South to North through Marotseland*, 2 vols. (London: John Lane, 1904), 2:33, 38–39, 44.

29. "Extract from Mr. Carlisle's letter of September 2nd, 1904," encl. in R. T. Coryndon to HCSA, Kalomo, 19 November 1904, NAZ, BSAC/NW/HC1/2/14.

30. E. A Copeman, "Extract from report . . . ," encl. in SNA to S. Admin., Kalomo, 23 November 1906, NAZ, BSAC/NW/A3/24/9. See also Pélissier, *Guerres grises*, 404n34. For the origins of the BNP, see chap. 6.

31. G. A. McGregor to L. A. Wallace, Livingstone, 13 September 1909, encl. in L. A. Wallace to HCSA, Livingstone, 5 November 1909, NAZ, BSAC/NW/HC1/2/43. The expression "policy of violence" is the Earl of Crewe's; Colonial Secretary to HCSA, London, 20 January 1910, NAZ, BSAC/NW/HC1/2/48. On local memories of "Mac," see James A. Pritchett, *Friends for Life, Friends for Death: Cohorts and Consciousness among the Lunda-Ndembu* (Charlottesville: University of Virginia Press, 2007), 48–51.

32. C. Bellis, Annual Report, "Kalulua Station," 25 April 1910, NAZ, KSE 6/1/1.

33. L. A. Wallace to HCSA, Livingstone, 21 January 1910, NAZ, BSAC/NW/HC1/2/48.

34. C. Harding to Foreign Secretary, Lealui, 16 January 1901, encl. in S. Admin. to Imperial Secretary (henceforth IS), Bulawayo, 30 April 1901, NAZ, BSAC/NW/HC1/2/1.

35. R. T. Coryndon to Secretary (BSAC), London, 1 March 1904, NAZ, BSAC/NW/A2/2/2.

36. C. Harding to IS, Kalomo, 16 May 1903, encl. in R. T. Coryndon to IS, Kalomo, 18 May 1903, NAZ, BSAC/NW/HC1/2/6.

37. R. T. Coryndon to Secretary (BSAC), London, 1 March 1904, NAZ, BSAC/NW/A2/2/2.

38. [F. C. Macaulay?] to [S. Admin.?], n.p., "April" 1903, NAZ, BSAC/NW/IN 2/1/11.

39. [F. C. Macaulay], "Report for Year 1902," n.p., 14 March 1903, NAZ, BSAC/NW/IN 2/1/11.

40. "Extract from Report of the Assistant District Commissioner, Kasempa, for the Month of October 1905," NAZ., BSAC/NW/A3/24/9.

41. See, e.g., C. Harding to Foreign Secretary, Lealui, 4 July 1900, NAZ, KDE 2/44/1–3; V. Gielgud to R. E. Codrington, Muyanga's, 21 November 1900, NAZ, BSAC/NER/A3/6/1; C. Harding to Foreign Secretary, Lealui, 16 January 1901, encl. in S. Admin. to IS, Bulawayo, 30 April 1901, NAZ, BSAC/NW/HC1/2/1

42. James Stevenson-Hamilton, *The Barotseland Journal of James Stevenson-Hamilton, 1898–1899*, ed. J. P. R. Wallis (London: Chatto & Windus, 1953), 225.

43. V. Gielgud to R. E. Codrington, Chipepo's, 25 September 1900, NAZ, BSAC/NER/A3/6/1. One year later, Maala, one of the largest and most influential communities in Bwila, would be "conquered" without the European forces having to confront any enemy fire. John K. Rennie, "The Conquest of Maala in 1901: An Exercise in Oral and Documentary Evidence among the Ila of Namwala District, Zambia," paper presented to the Institute for African

Studies and the Department of History, University of Zambia, Lusaka, February 1982.

44. V. Gielgud to R. E. Codrington, Muyanga's, 14 October 1900, NAZ, BSAC/NER/A3/6/1.

45. V. Gielgud to R. E. Codrington, Muyanga's, 11 October 1900, NAZ, BSAC/NER/A3/6/1. Long excerpts from this report have been published in S. R. Denny, "Val Gielgud and the Slave Traders," *Northern Rhodesia Journal* 3, no. 4 (1957): 331–32.

46. V. Gielgud to R. E. Codrington, Muyanga's, 14 October 1900, NAZ, BSAC/NER/A3/6/1.

47. The relationship between gun legislation and ideas about citizenship in colonial South Africa is the central theme of William K. Storey, *Guns, Race, and Power in Colonial South Africa* (Cambridge: Cambridge University Press, 2008).

48. W. Hazell to L. A. Wallace, 25 March 1912, encl. in L. A. Wallace to H. J. Gladstone, 30 March 1912, NAZ, NR/HC1/3/1, vol. 2.

49. Ibid.

50. L. A. Wallace to H. J. Gladstone, 30 March 1912, NAZ, NR/HC1/3/1, vol. 2.

51. H. J. Gladstone, "Affairs in the Kasempa District—Northern Rhodesia," 18 April 1912, encl. in IS to L. A. Wallace, Cape Town, 20 April 1912, NAZ, BSAC/NR/HC1/3/1, vol. 2.

52. Macdonnel to S. Admin., 20 June 1912. The word "rifle" in the quotation must be taken to be an ill-chosen synonym for "gun."

53. SNA, "Notes on the Draft Arms and Ammunition Proclamation," Livingstone, 2 July 1912, NAZ, BSAC/NR/ B1/2/368.

54. Proclamation no. 9 of 1912, in *Northern Rhodesia Government Gazette* 2, no. 8 (1 August 1912); L. Harcourt to H. J. Gladstone, 8 June 1912, NAZ, BSAC/NR/A1/1/5.

55. Proclamation no. 18 of 1914, in *Northern Rhodesia Government Gazette* 4, 16 (29 August 1914).

56. Ibid., and SNA to All District Officials, Livingstone, 18 January 1923, NAZ, BSAC/NR/B1/2/368.

57. SNA to DC (Kasempa), [Livingstone], 18 June 1923, NAZ, ZA 1/9/51/2.

58. C. S. Parsons to DC (Kasempa), Solwezi, 23 January 1923, NAZ, BSAC/NR/B1/2/368.

59. P. Hall to SNA, Kasempa, 23 January 1923, NAZ, BSAC/NR/B1/2/368.

60. Ibid.

61. R. E. B. Woods to DC (Kasempa), n.p., 25 January 1923, NAZ, BSAC/NR/B1/2/368.

62. Ibid.

63. Ibid.

64. Yeta et al. to Resident Magistrate (Barotseland), Lealui, 31 March 1923, encl. in Resident Magistrate to SNA, Mongu, 4 April 1923, NAZ, BSAC/NR/B1/2/368.

65. F. V. Bruce Miller to Resident Magistrate (Barotseland), Balovale, 4 September 1923, NAZ, ZA1/9/51/2.

66. SNA to All District Officials, Livingstone, 18 January 1923, NAZ, BSAC/NR/B1/2/368. See also SNA to Legal Adviser, Livingstone, 19 November 1923, ZA1/9/51/2.

67. K. S. Kinross, "Kasempa Sub-District: Annual Report for the Year Ending 31 March 1924," NAZ, ZA7/1/7/6.

68. P. Hall, "Kasempa District: Annual Report for the Year Ending 31 March 1924," NAZ, ZA7/1/7/6.

69. P. Hall to Attorney-General, Kasempa, 21 September 1924, NAZ, ZA 1/9/51/2.

70. SNA to P. Hall, [Livingstone], 4 March 1924; and P. Hall, "Firearms Restriction Proclamation 21/1922: Supplementary Instructions," encl. in P. Hall to SNA, Kasempa, 13 March 1924, both in NAZ, ZA1/9/51/2.

71. C. Rennie to SNA, Kasempa, 22 April 1926, ZA 1/9/51/2. See also Criminal Case 22 of 1925, Mwinilunga, 11 Feb. 1925, and Criminal Case 43 of 1925, Mwinilunga, 25 March 1925, both in NAZ, KSE 3/2/2/3–4.

72. Rennie to SNA, 22 April 1926.

73. Ibid.

74. I. A. Bartling, "A Living Church Cannot Stop at Home," *South Africa General Mission Pioneer* 37, no. 12 (December 1924): 148.

75. Ordinance no. 19 of 1925, in *Northern Rhodesia Government Gazette* 15, no. 12 (13 June 1925).

76. Writing in the 1970s, ethnographer Han Bantje remarked that Kaonde hunting had been in decline for the past "half century." (Han Bantje, *Kaonde Song and Ritual*, Annales du musée royal de l'Afrique centrale 95 [Tervuren, Belgium: Musée royal de l'Afrique centrale, 1978], 7). Ten or so years later, however, elements of the old Kaonde hunting and gun culture reemerged to the surface in the context of Zambia's then desperate economic crisis and waning state power. Kate Crehan, *The Fractured Community: Landscapes of Power and Gender in Rural Zambia* (Berkeley: University of California Press, 1997), chap. 5.

77. M. Rutten to Public Prosecutor, Lukafu, n.d. [late March–early April 1905], MRAC, Léon Guebels's Papers (GP), RG 915; D. Campbell to H. R. Fox Bourne, Johnston Falls, 14 May 1904, in Edmund D. Morel, *King Leopold's Rule in Africa* (London: William Heinemann, 1904), 457–58. Dugald Campbell's important letter to Henry R. Fox Bourne, the secretary of the Aborigines Protection Society, is also to be found in *The Aborigines' Friend* (October 1904): 201–14, and in NAUK, FO10/811. See also Vellut, "Garenganze/Katanga–Bié–Benguela," 138–39, 147.

78. Mr. Heide, "Rapport sur le commerce en fraude d'armes et munitions," Kongolo, 27 July 1911, AA, AE (309) 415.

79. C. Brasseur to D. Brasseur, Lofoi, "1 January 1895" [sic, but 12 January 1895], MRAC, BP, and D. Crawford, 1 August 1895, *Echoes of Service* 350 (January 1986): 26.

80. C. Brasseur to D. Brasseur, Lofoi, 25 November 1895, MRAC, BP.

81. D. Crawford, 5 January 1896, *Echoes of Service* 364 (August 1896): 250. See also D. Campbell, 13 March 1897, *Echoes of Service* 390 (September 1897): 283, and M. Rutten to Public Prosecutor, Lukafu, 24 October 1902, MRAC, GP.

82. For Brasseur's military background, see Institut Royal Colonial Belge, *Biographie Coloniale Belge*, vol. 1 (Brussels: Librairie Falk Fils, 1948), 162–63.

83. The first and third quotes are to be found in D. Crawford, 25 September 1893, *Echoes of Service* 308 (April 1894): 101–2; the second ("je considère tous les nègres comme des juifs du premier ordre . . . n'obéissant qu'à la force") in C. Brasseur to D. Brasseur, Lofoi, "20 October 1894" [*sic*, but between 23 and 31 October 1894], MRAC, BP.

84. Brasseur admitted to his brother that he expected "*un certain %*" out of the 1,600–1,700 kg of ivory that he had obtained during his first fifteen or so months in southern Katanga; Brasseur to Brasseur, "1 January 1895."

85. "[U]n oiseau dont la face interne des ailes est d'un rouge sanglant. Or, disent les indigènes, Mr. Brasseur n'était content que quand il avait du sang jusqu'aux aisselles. Alors il ressemblait à l'oiseau en question." J.-M. Jenniges to Public Prosecutor, Lukafu, 1 March "1903" [*sic*, but 1905], MRAC, GP.

86. Campbell to Fox Bourne, in Morel, *King Leopold's Rule*, 456. The Brethren had been based near the Lofoi station since the great Bunkeya famine of 1891.

87. D. Crawford, 1 December 1895, *Echoes of Service* 357 (May 1896): 135.

88. Campbell to Fox Bourne, in Morel, *King Leopold's Rule*, 460. See also, e.g., D. Campbell, 7 May 1895, *Echoes of Service* 345 (November 1895): 268, and D. Crawford, 1 September 1895, *Echoes of Service* 351 (February 1896): 45–46.

89. "Tu étais pauvre. Tu te vantes même dans la région. . . . C'est parce que les Belges sont à Bunkeya." Léon Verbeek, *L'histoire dans les chants et les danses populaires: La zone culturelle Bemba du Haut-Shaba (Zaïre)*, Enquêtes et documents d'histoire Africaine, 10 (Louvain-la-Neuve, Belgium: Centre d'histoire de l'Afrique, 1992), 81–82.

90. Brasseur to Brasseur, "20 October 1894," and C. Brasseur to D. Brasseur, Lofoi, 12 March 1897, MRAC, BP.

91. D. Campbell, 2 May 1897, *Echoes of Service* 397 (January 1898): 13.

92. "[U]ne bande de loups affamés se jetant sur un cadavre." Brasseur to Brasseur, "20 October 1894."

93. Ibid.

94. Ibid., and Brasseur to Brasseur, 12 March 1897.

95. Clément Brasseur, "Lettre sur le Katanga," *La Belgique Coloniale* 2, no. 17 (1896): 197; Paul Le Marinel, *Carnets de route dans l'État Indépendant du Congo de 1887 à 1910* (Brussels: Éditions Progress, 1991), 206.

96. Campbell to Fox Bourne, in Morel, *King Leopold's Rule*, 460.

97. Campbell to Fox Bourne, ibid., 454.

98. Campbell to Fox Bourne, ibid., 456. Reference to the same episode is made in Campbell, *In the Heart of Bantuland*, 65–66, where the figure of twenty-two tusks is given.

99. Campbell to Fox Bourne, in Morel, *King Leopold's Rule*, 456.

100. "[T]oujours accompagné de nombreux auxiliaires de race bayeke." M. Rutten to Public Prosecutor, Lukafu, 6 March, 1905, MRAC, GP.

101. Rutten to Public Prosecutor, n.d., MRAC, GP.

102. W. George, 3 January 1898, *Echoes of Service* 409 (July 1898): 204.

103. See, e.g., "Notes of a Journey from Mwena, Garenganze, to Kavungu, Lovale Country," *Echoes of Service* 403 (April 1898): 111, and D. Crawford, 14 June 1899, *Echoes of Service* 441 (November 1899): 333.

104. W. George, 12 August 1899, *Echoes of Service* 448 (February 1900): 57. See also Pierre Kalenga, "Situation socio-politique du Katanga ancien (1870–1911)" (mémoire de maîtrise, University of Lubumbashi, DR Congo, 2010), 119–20.

105. "Acte Général de la Conférence de Bruxelles sur la Traite des Esclaves et le Régime de Spiritueux en Afrique," 2 July 1890, in *Lois en vigueur dans l'État Indépendant du Congo*, comp. Octave Louwers (Brussels: Weissenbruch, 1905), 60. Gun licenses for flintlock muzzle-loaders were only introduced in 1912; "Décret—Armes à Feu et Munitions," 6 January 1912, *Bulletin Officiel du Congo Belge* 5, no. 1 (15 January 1912).

106. "Note," n.d., encl. in Directeur Général (2e Direction Générale, Ministère des Colonies) to Directeur Général (1e Direction Générale, Ministère des Colonies), Brussels, 18 January 1911, AA, AE (307) 415.

107. The case of the Songye warlord, Lumpungu, who was apparently given as many as a thousand percussion muskets by the CFS in the early 1890s, is discussed by Donatien Dibwe dia Mwembu, "The Role of Firearms in the Songye Region, 1869–1960," in *The Objects of Life in Central Africa: The History of Consumption and Social Change, 1840–1980*, ed. Robert Ross, Marja Hinfelaar, and Iva Peša (Leiden: Brill, 2013), 41–64. Insightful general remarks about the centrality of African collaboration to the early history of the CFS are to be found in both Jean-Luc Vellut, "La violence armée dans l'État Indépendant du Congo: Ténèbres et clartés dans l'histoire d'un état conquérant," *Cultures et Développement* 16, nos. 3–4 (1984): 671–707, and Aldwin Roes, "Towards a History of Mass Violence in the Etat Indépendant du Congo, 1885–1908," *South African Historical Journal* 62, no. 4 (2010): 634–70.

108. C. Brasseur to D. Brasseur, Lofoi, "1 May 1895" [*sic*, but 9 October 1895], MRAC, BP.

109. Clément Brasseur, "Lettre sur le Katanga," *La Belgique Coloniale* 3, no. 20 (1897): 234.

110. Campbell to Fox Bourne, in Morel, *King Leopold's Rule*, 455.

111. "État récapitulatif des armes à feu autres que les fusils à silex non rayés pour lesquelles des permis de port d'armes ont été délivrés . . . pendant le mois de Novembre 1905," AA, AE (302) 408.

112. Campbell to Fox Bourne, in Morel, *King Leopold's Rule*, 453.

113. D. Crawford, 24 February 1892, *Echoes of Service* 274 (November 1892): 269.

114. D. Crawford, 5 January 1896, *Echoes of Service* 364 (August 1896): 250.

115. D. Campbell, 22 December 1896, *Echoes of Service* 386 (July 1897): 218–19.

116. D. Campbell, 6 June 1897, *Echoes of Service* 394 (November 1897): 345.

117. Campbell to Fox Bourne, in Morel, *King Leopold's Rule*, 460.

118. L. George, 1 November 1899, *Echoes of Service* 452 (April 1900): 122.

119. "Les environs de Lofoi et de Lukafu sont peuplés de femmes jadis faites prisonnières à la guerre." Rutten to Public Prosecutor, Lukafu, 6 March 1905.

120. Campbell to Fox Bourne, in Morel, *King Leopold's Rule*, 457, 460.

121. "[À] chacun aussi j'avais donné une femme." C. Brasseur to D. Brasseur, Lofoi, "1 April 1895."

122. Jenniges to Public Prosecutor, Lukafu, 1 March "1903."

123. Chef de Poste, "Rapport mensuel général sur la situation du poste," Lukafu, 30 April 1909, Fiche 1582, and Chef de Zone du Haut-Luapula to Vice-Governor General of Katanga, Lukafu, 16 September 1910, Fiches 1586/7, both in "Documents du Zaïre Colonial—Collection J. L. Vellut," Bibliothèque africaine at the Bibliothèque des Affaires étrangères, Brussels, Belgium. Lutipisha was abandoned following the promulgation of anti–sleeping sickness regulations.

124. Author's interviews with J. Mwidye Kidyamba, Bunkeya, 5 August 2011, and B. Mwenda Numbi, Bunkeya, 7 August 2011.

125. "Rapport du Conseil Colonial sur le projet de décret réglant les droits de chasse et de pêche," 16 July 1910, and "Décret du Roi-Souverain—Droits de chasse et de pêche au Congo belge," 26 July 1910, both in *Bulletin Officiel du Congo Belge* 3, no. 13 (29 July 1910): 638–52.

126. Brasseur to Brasseur, "1 April 1895"; Brasseur to Brasseur, 12 March 1897.

127. D. Campbell, 6 June 1897, *Echoes of Service* 394 (November 1897): 345.

128. D. Crawford, 13 December 1902, *Echoes of Service* 526 (May 1903). See also Campbell, *In the Heart of Bantuland*, 187.

129. Campbell, *In the Heart of Bantuland*, 191.

130. G. Cuvelier, Annual Report for the Territoire de Jadotville, Jadotville, 31 January 1933, AA, RA/AIMO 95. The right of Congolese Africans to own percussion-lock muskets had been enshrined in law in 1915; "Ordonnance du Gouverneur Général—Armes à feu et leurs munitions," 31 August 1915, in *Codes et lois du Congo Belge*, comp. Octave Louwers and Iwan Grenade, with C. Kuck (Brussels: Weissenbruch, 1927), 761.

131. Interview with J. Mwidye Kidyamba.

132. Author's interview with Nshita Kafuku, Mwami Mukonki VIII, Bunkeya, 8 August 2011.

133. Ibid.

134. Chef de Poste, "Rapport mensuel," 30 April 1909.

135. Léon Verbeek, *Le chasseur africain et son monde: Chansons de chasse du sud-est du Katanga* (Tervuren: Musée royal de l'Afrique centrale, 2007), 34, http://www.africamuseum.be/research/publications/rmca/online/chansons%20de%20chasse.pdf.

136. "Toi, mon fusil, tu m'as fait honte / Dans cette brousse . . . il y a des animaux, papa / Il y a des animaux, mon fusil, tu m'as fait honte." Ibid., 95–97.

137. Ibid., 94.

138. Gendarmes from Bunkeya are known to have devised a special salute to greet one another; interview with B. Mwenda Numbi. See also Erik Kennes, "Fin du cycle post-colonial au Katanga, RD Congo: Rébellions, sécession et leurs mémoires dans la dynamique des articulations entre l'État central et l'autonomie régionale, 1960–2007" (PhD diss., Université Laval and Université Paris I, 2009), 419.

CHAPTER 5: "THEY DISDAIN FIREARMS"

1. Norman Etherington, *The Great Treks: The Transformation of Southern Africa, 1815–1854* (Harlow: Longman, 2001), 114–21, 275–77.

2. John A. Barnes, *Politics in a Changing Society: A Political History of the Fort Jameson Ngoni* (Cape Town: Oxford University Press, 1954).

3. Carl Wiese, *Expedition in East-Central Africa, 1888–1891: A Report*, ed. Harry W. Langworthy, trans. Donald Ramos (Norman: University of Oklahoma Press, 1983), 160n41.

4. T. Jack Thompson, "The Origins, Migration, and Settlement of the Northern Ngoni," *Society of Malawi Journal* 34, no. 1 (1981): 16.

5. Etherington, *Great Treks*, 277.

6. Accounts of Ngoni migrations are to be found in several sources. The above summary draws mainly on those of J. P. Bruwer, *Mbiri ya Angoni* (Mkhoma, Malawi: DRC Mission Press, 1941); Ian Linden, "Some Oral Traditions from the Maseko Ngoni," *Society of Malawi Journal* 24, no. 2 (1971): 60–73; William E. Rau, "Mpezeni's Ngoni of Eastern Zambia, 1870–1920" (PhD diss., UCLA, 1974), chap. 2; Andrew D. Roberts, *A History of Zambia* (London: Heinemann, 1976), 118–20; and Thompson, "Origins, Migration, and Settlement."

7. See, e.g., Robert Laws, "Journey along Part of the Western Side of Lake Nyassa, in 1878," *Proceedings of the Royal Geographical Society* 1, no. 5 (1879): 321.

8. The first and the third quotes are to be found in John McCracken, *Politics and Christianity in Malawi, 1875–1940: The Impact of the Livingstonia Mission in the Northern Province* (Cambridge: Cambridge University Press, 1977), 8; the second in Rau, "Mpezeni's Ngoni," 171.

9. Leroy Vail, "The Making of the 'Dead North': A Study of the Ngoni Rule in Northern Malawi, c. 1855–1907," in *Before and After Shaka: Papers in Nguni History*, ed. Jeffrey B. Peires (Grahamstown, South Africa: Institute of Social and Economic Research, Rhodes University, 1981), 230.

10. W. A. Elmslie, *Among the Wild Ngoni* (Edinburgh: Oliphant, Anderson & Ferrier, 1899), 50, 78, 89.

11. My overall approach is influenced by Vail, "Making of the 'Dead North,'" an important revisionist account which, however, errs in the opposite direction, painting too gloomy a picture of the condition of M'Mbelwa's kingdom on the eve of colonialism.

12. See, e.g., Giacomo Macola, "The History of the Eastern Lunda Royal Capitals to 1900," in *The Urban Experience in Eastern Africa, c. 1750–2000*, ed. Andrew Burton, special issue, *Azania* 14, nos. 36–37 (2002): 31–45, and Richard J. Reid, *War in Pre-Colonial Eastern Africa: The Patterns & Meanings of State-Level Conflict in the Nineteenth Century* (Oxford: James Currey, 2007), 61–62.

13. Mwase Kasungu's victory may have preceded Livingstone's visit to Kasungu in September 1863. "Muazi"—the Scottish explorer reported—"ha[d] suffered from the attacks of the Mazitu [Ngoni]," but he had "evidently clung to his birthplace." David Livingstone and Charles Livingstone, *Narrative of an Expedition to the Zambezi and its Tributaries* (London: John Murray, 1865), 526.

14. T. Cullen Young, *Notes on the History of the Tumbuka-Kamanga Peoples in the Northern Province of Nyasaland* (1932; repr., London: Frank Cass, 1970), 125.

15. Ibid.; McCracken, *Politics and Christianity*, 12.

16. Kings M. Phiri, "Chewa History in Central Malawi and the Use of Oral Tradition, 1600–1920" (PhD diss., University of Wisconsin–Madison, 1975), 157–58. The same author, however, dates Mwase Kasungu's victory against the Ngoni to the 1870s.

17. Ibid., 158.

18. Andrew D. Roberts, *A History of the Bemba: Political Growth and Change in North-Eastern Zambia before 1900* (London: Longman, 1973), 119–21, 128, 142–45.

19. Ibid., 203.

20. Harry W. Langworthy, "Swahili Influence in the Area between Lake Malawi and the Luangwa River," *African Historical Studies* 4, no. 3 (1971): 585–86.

21. Donald Fraser, *Winning a Primitive People: Sixteen Years' Work among the Warlike Tribe of the Ngoni and the Senga and Tumbuka Peoples of Central Africa* (1914; repr., London: Seeley, Service, 1922), 85.

22. Roberts, *History of the Bemba*, 374–75.

23. W. A. Elmslie to R. Laws, "Angoniland," 10 December 1887, National Library of Scotland (NLS), Edinburgh, MS 7890; McCracken, *Politics and Christianity*, 87.

24. McCracken, *Politics and Christianity*, 86. McCracken's main original source, the Bandawe Mission Journal of 1885 (NLS, MS 7911), was not available for consultation at the time of my visit to the NLS.

25. Despite some scholarly confusion surrounding its timing, the onset of the Tonga rebellion can be dated fairly precisely to the period between June

1877, when H. B. Cotterill spent a few days among the Tonga of Mankham-bira and saw no sign of ongoing hostilities ("On the Nyassa and a Journey from the North End to Zanzibar," *Proceedings of the Royal Geographical Society* 22, no. 4 [1877–78]: 233–51), and November of the same year, when Dr. James Stewart reported on the large number of temporary huts recently built at Mankhambira's "by people driven in from surrounding villages by the war with the Maviti." ("The Second Circumnavigation of Lake Nyassa," *Proceedings of the Royal Geographical Society* 1, no. 5 [1879]: 299). See also Thompson, "Origins, Migration, and Settlement," 35n114.

26. The first quotation is to be found in Stewart, "Second Circumnavigation," 299; the second in Robert Laws, "Journey along Part," 314.

27. Stewart, "Second Circumnavigation," 300.

28. Jaap van Velsen, "Notes on the History of the Lakeside Tonga of Nyasaland," *African Studies* 18, no. 3 (1959): 116.

29. Yesaya M. Chibambo, *My Ngoni of Nyasaland*, trans. Charles Stuart (London: United Society for Christian Literature, n.d. [1942]), 43.

30. Robert Laws's travel diary, 7 October 1878, Edinburgh University Library, Centre for Research Collections, Edinburgh, Papers of Robert Laws, Gen 561/1. Laws did not include the above account in his published "Journey."

31. Bandawe Mission Journal, 30 November 1878, 10 and 26 January 1879, NLS, MS 7910; McCracken, *Politics and Christianity*, 60–61.

32. Bandawe Mission Journal, 6 December 1881, quoted in Elmslie, *Among the Wild Ngoni*, 101.

33. L. Goodrich to Secretary of State for Foreign Affairs, Bandawe, 19 February 1885, NAUK, FO 84/1702.

34. See, e.g., W. P. Livingstone, *Laws of Livingstonia: A Narrative of Missionary Adventure and Achievement* (London: Hodder and Stoughton, 1921), 195, 204, 210.

35. Elmslie, *Among the Wild Ngoni*, 101.

36. Elmslie to Laws, 10 December 1887; W. A. Elmslie to R. Laws, "Angoniland," 13 May 1888, NLS, MS. 7891.

37. See, e.g., Laws's travel diary, 25–26 September 1878 (slightly embellished in Livingstone, *Laws of Livingstonia*, 156), and Frederick L. M. Moir, *After Livingstone: An African Trade Romance* (London: Hodder and Stoughton, n.d. [1923]), 27–28.

38. L. Goodrich to Secretary of State for Foreign Affairs, "Angoniland," 24 April 1885, NAUK, FO 84/1702.

39. Laws's travel diary, 10 October 1878.

40. McCracken, *Politics and Christianity*, 57.

41. Kaningina Mission Journal, 7 and 15 February 1879, 11 March 1879, 1 and 28 April 1879, NLS, MS 7910.

42. Bandawe Mission Journal, 14 May 1879.

43. Ibid.

44. Elmslie to Laws, 10 December 1887. See also Elmslie, *Among the Wild Ngoni*, 144.

45. W. A. Elmslie to R. Laws, "Angoniland," 22 March 1888, NLS, MS. 7891.

46. Frederick D. Lugard, *The Rise of Our East African Empire*, 2 vols. (Edinburgh: William Blackwood, 1893), 1:86–87.

47. Laws, "Journey along Part," 308.

48. Ibid.; Laws's travel diary, 10 August 1878.

49. Laws's travel diary, 20 August 1878.

50. Edward D. Young, *Nyassa: A Journal of Adventures whilst Exploring Lake Nyassa, Central Africa, and Establishing the Settlement of "Livingstonia"* (London: John Murray, 1877), 179.

51. Walter M. Kerr, *The Far Interior: A Narrative of Travel and Adventure from the Cape of Good Hope across the Zambesi to the Lake Regions of Central Africa*, 2 vols. (Boston: Houghton, Mifflin, 1886), 2:58.

52. Ibid., 2:126.

53. Ibid, 2:124.

54. See, e.g., the descriptions included in Laws's travel diary, 15 August 1878, and J. T. Last, "A Journey from Blantyre to Angoni-Land and Back," *Proceedings of the Royal Geographical Society* 9, no. 3 (1887): 186.

55. Kerr, *Far Interior*, 2:125.

56. Though not entirely absent; see ibid., 2:70–71, 112.

57. Ibid., 2:89–90.

58. Ibid., 2:154, 178.

59. Ibid., 2:106, 131.

60. L. Goodrich to Secretary of State for Foreign Affairs, "Angoniland," 19 March 1885, NAUK, FO 84/1702.

61. Last, "Journey from Blantyre to Angoni-Land," 180.

62. A. G. S. Hawes to Secretary of State for Foreign Affairs, Mandala, 7 July 1886, published in House of Commons, *Accounts and Papers*, 78 (1887) [C.5111].

63. A. G. S. Hawes to Secretary of State for Foreign Affairs, Livingstonia, 3 June 1886, published in House of Commons, *Accounts and Papers*, 78 (1887) [C.5111].

64. Diary of F. J. Morrison, an employee of the African Lakes Company, March 1883, quoted in Andrew C. Ross, *Blantyre Mission and the Making of Modern Malawi* (Blantyre, Malawi: Christian Literature Association in Malawi, 1996), 77. Morrison's remarks referred to the arsenal of Mponda I, who died early in 1886 and whom another source described as having owned "guns by the hundreds." Ian Linden, ed. and trans., "Mponda Mission Diary, 1889–1891: Daily Life in a Machinga Village: Part II," *International Journal of African Historical Studies* 7, no. 3 (1974): 501.

65. Hawes to Secretary of State, 7 July 1886.

66. Last, "Journey from Blantyre to Angoni-Land," 186.

67. Young, *Nyassa: A Journal of Adventures*, 176.

68. Ibid., 174–75.

69. John Buchanan, *The Shirè Highlands (East Central Africa) as Colony and Mission* (Edinburgh: William Blackwood, 1885), 94. See also Landeg

White, *Magomero: Portrait of an African Village* (Cambridge: Cambridge University Press, 1987), 65.

70. Ian Linden, ed. and trans., "Mponda Mission Diary, 1889–1891: Daily Life in a Machinga Village: Part I," *International Journal of African Historical Studies* 7, no. 2 (1974): 277.

71. Ian Linden, "The Maseko Ngoni at Domwe, 1870–1900," in *The Early History of Malawi*, ed. Bridglal Pachai (London: Longman, 1972), 241.

72. See, e.g., Linden, "Mponda Mission Diary, 1889–1891: Daily Life in a Machinga Village: Part II," 499.

73. Linden, "Mponda Mission Diary, 1889–1891: Daily Life in a Machinga Village: Part I," 302.

74. Ian Linden, ed. and trans., "Mponda Mission Diary, 1889–1891: Part IV: The Ngoni Defeat and Missionary Departure," *International Journal of African Historical Studies* 8, no. 1 (1975): 116.

75. Ibid., 117. Besides muzzle-loaders and Enfield rifles, Mponda II's thousand guns may have also included some Winchester repeating rifles; William P. Johnson, *My African Reminiscences, 1875–1895* (London: Universities' Mission to Central Africa, n.d. [1924?]), 164.

76. Linden, "Mponda Mission Diary: Part IV," 117. A slightly different, but consistent, rendering of the French original is given in Linden, "The Maseko Ngoni at Domwe," 241.

77. Linden, "Mponda Mission Diary: Part IV," 123.

78. In light of the development of the battle described above, and countless other contrasting remarks in the same missionary source, the report that the Yao marauders had found an "enormous quantity of guns" at Chikusi's is probably to be viewed as either fabrication or exaggeration; Linden, "Mponda Mission Diary: Part IV," 131n29.

79. Livingstone and Livingstone, *Narrative of an Expedition*, 502; and Andrew D. Roberts, "Firearms in North-Eastern Zambia before 1900," *Transafrican Journal of History* 1, no. 2 (1971): 5, where "1861" should read "1863."

80. See, e.g., the case of chief Msoro, as discussed by Albert J. Williams-Myers, "The Nsenga of Central Africa: Political and Economic Aspects of Clan History, 1700 to the Late Nineteenth Century" (PhD diss., UCLA, 1978), 293–94.

81. Ibid., 266; Allen F. Isaacman and Barbara S. Isaacman, *Slavery and Beyond: The Making of Men and Chikunda Ethnic Identities in the Unstable World of South-Central Africa, 1750–1920* (Portsmouth, NH: Heinemann, 2004), 205.

82. Wiese, *Expedition in East-Central Africa*, 275.

83. Ibid., 168.

84. Ibid., 244, 252, 254. Chibisa was especially impressed by the rifle of Lieutenant Solla, Wiese's companion. This was a "Kropaczeck," that is, a Portuguese Mauser-Kropatschek M1886/89, which one present-day dealer describes as "perhaps one of the finest, most well made rifles of its generation, with an incredibly smooth action throughout"; "Portuguese Mauser-Kropatschek

M1886/89," last modified 26 January 2010, accessed 30 January 2011, http://www.antiquemilitaryrifles.com/showrifle.asp?r=669.

85. Wiese, *Expedition in East-Central Africa*, 168.

86. Alfred Sharpe, "A Journey through the Country Lying between the Shire and Loangwa Rivers," *Proceedings of the Royal Geographical Society* 12, no. 3 (1890): 156; Wiese, *Expedition in East-Central Africa*, 113, 117, 131–32.

87. "Les gens de Mpéseni . . . sont pourtant encore souvent battus, n'étant armés que d'arcs, de boucliers, de sagaies et de casse-tête, tandis que leurs adversaires, les Agoas ou gens d'Oundi, ont des fusils et de la poudre." Édouard Foà, *Du Cap au Lac Nyassa* (1897; repr., Paris: Librairie Plon, 1901), 280.

88. Wiese, *Expedition in East-Central Africa*, 179.

89. Harry W. Langworthy, "Introduction: Carl Wiese and Zambezia," in Wiese, *Expedition in East-Central Africa, 1888–1891: A Report*, ed. Harry W. Langworthy (Norman: University of Oklahoma Press, 1983), 4–5.

90. Wiese, *Expedition in East-Central Africa*, 139, 195.

91. Ibid., 155, 186.

92. Ibid., 185.

93. Ibid., 191, 165.

94. Ibid., 185–86.

95. Ibid., 191.

96. Ibid., 152.

97. Ibid., 160.

98. Ibid., 195. Mpezeni's collection, Wiese went on, included "an abundance of different revolvers, Snider, Spencer, Remington, Lefaucheux, and other carbines, as well as huge amounts of different percussion arms, some presented to him, others taken in wars."

99. Ibid., 153–54. See also Carl Wiese, "Beiträge zur geschichte der Zulu im norden des Zambesi, namentlich der Angoni," *Zeitschrift für Ethnologie* 32 (1900): 196, where Wiese speaks explicitly of "avoidance" (vermeidung) of firearms on the part of Mpezeni's Ngoni. "[E]ven today"—he elaborated in the same article—"the spear, shield and club are their only weapons, just as in the time of their fathers before they crossed the Zambezi. Although they amass firearms from enemies during battles, these would only be used to fire salute shots at festivities." ([N]och heute sind Speer, Schild und Keule ihre einzige Bewaffnung, ganz so wie in der Zeit ihrer Väter, ehe sie den Zambesi überschritten. Sie haben in ihren Kriegszügen wohl eine Menge Feuerwaffen von den anderen Stämmen erbeutet, benützen dieselben jedoch nur, um bei Festlichkeiten und Gelagen Salutschüsse abzugeben.) My thanks to Judith Weik for her translations from German.

100. Wiese, *Expedition in East-Central Africa*, 157.

101. R. G. Warton to P. Forbes, [Beira?], 2 April 1896, in *North Charterland Concession Inquiry: Report to the Governor of Northern Rhodesia by the Commissioner, Mr. Justice Maugham, July 1932* (London: HMSO, 1932), appendix 28.

102. R. G. Warton to Secretary (NCEC), Luangeni, 6 August 1896, in "Copy by E. H. Lane Poole of correspondence relating to the North

Charterland Exploration Company (East Loangwa District), 1896. Also relating to the Angoni Rising" (hereafter Poole Papers), Archives of the Livingstone Museum, no ref., Livingstone, Zambia. Poole's "Copy" consists of verbatim copies and excerpts of documents that the author may have consulted in the Eastern Province, where he served as a colonial administrator in the 1920s, in preparation for his "A short history of the North Charterland Exploration Company," Lundazi, 3 February 1923, Archives of the Livingstone Museum, LM2/4/31. This latter text was published several years later as E. H. Lane Poole, "Mpeseni and the Exploration Companies, 1885–1898," *Northern Rhodesia Journal* 5, no. 3 (1963): 221–32. My thanks to Jack Hogan for making copies of the Poole Papers available to me.

103. Margaret Read, "Tradition and Prestige among the Ngoni," *Africa: Journal of the International African Institute* 9, no. 4 (1936): 461m. See Barnes, *Politics in a Changing Society*, 59; Vail, "Making of the 'Dead North,'" 249; and, most recently, Michael W. Conner, *The Art of the Jere and Maseko Ngoni of Malawi, 1818–1964* (New York: Man's Heritage Press, 1993). Read's original source was probably Young, *Nyassa: A Journal of Adventures*, 176.

104. Barnes, *Politics in a Changing Society*, 36–37, 40; Roberts, *History of Zambia*, 119. For similar dynamics among the followers of Mzilikazi, see Etherington, *Great Treks*, 165–66.

105. Quoted in Margaret Read, *The Ngoni of Nyasaland* (1956; repr., London: Frank Cass, 1970), 29.

106. Fraser, *Winning a Primitive People*, 32.

107. Wiese, *Expedition in East-Central Africa*, 152.

108. Barnes, *Politics in a Changing Society*, 30–32, 37.

109. Fraser, *Winning a Primitive People*, 51. M'Mbelwa himself related the story of Ng'onomo to Acting Consul Goodrich with a view to emphasizing his opposition to the exportation of slaves and the incorporative nature of the polity he led; Goodrich to Secretary of State, 24 April 1885.

110. Wiese, *Expedition in East-Central Africa*, 155, and Chibambo, in Read, *Ngoni of Nyasaland*, 36–37.

111. Fraser, *Winning a Primitive People*, 37–39.

112. Chibambo, in Read, *Ngoni of Nyasaland*, 35.

113. Margaret Read's Papers, Archives of the London School of Economics, London, READ/1/10.

114. Margaret Read, "Songs of the Ngoni People," *Bantu Studies* 11, no. 1 (1937): 31.

115. Read, *Ngoni of Nyasaland*, 46.

116. Ibid.

117. This was also true of the Songea Ngoni of Tanzania; Heike Schmidt, "'Deadly Silence Predominates in this District': The Maji Maji War and Its Aftermath in Ungoni," in *Maji Maji: Lifting the Fog of War*, ed. James Giblin and Jamie Monson (Leiden: Brill, 2010), esp. 188–89.

118. John Iliffe, *Honour in African History* (Cambridge: Cambridge University Press, 2005), 4–6.

119. Last, "Journey from Blantyre to Angoni-Land," 186. In 1900, Wiese noted that the "spirit of the Zulu has been lost [by the Ngoni of Mpezeni], and their war tactics are mainly based on surprising the enemy. Fighting man against man, in which the old short Zulu spear was the main weapon of choice, has become a rarity." (Trotz alledem hat sich der alte unbändige Zulu-Geist verloren, und ihre ganze Kampfesweise besteht einfach in Ueberraschungen. Kämpfe Mann gegen Mann, wobei der alte kurze Zulu-Speer die Hauptrolle spielte, sind zu Seltenheiten geworden.) Wiese, "Beiträge zur geschichte," 196. Perhaps not too much emphasis ought to be placed on these remarks, which do not appear in Wiese's earlier travel diary and which are belied by the tactics employed by the Ngoni against the British early in 1898 (see chapter 6).

120. David M. Gordon, "The Abolition of the Slave Trade and the Transformation of the South-Central African Interior during the Nineteenth Century," *William and Mary Quarterly*, 3rd ser., 66, no. 4 (2009): 929.

121. These were the words of Mangwanana Mchunu, of the uVe, a regiment inaugurated by the Zulu king Cetshwayo c. 1875. Quoted in Ian Knight, *The Anatomy of the Zulu Army: From Shaka to Cetshwayo, 1818–1879* (London: Greenhill, 1995), 215, 268. On Buganda, see Richard J. Reid, *Political Power in Pre-Colonial Buganda: Economy, Society and Warfare in the Nineteenth Century* (Oxford: James Currey, 2002), 225.

122. Édouard Foà, *Du Cap au Lac Nyassa*, 298, and, by the same author, *La traversée de l'Afrique du Zambèze au Congo français* (Paris: Librairie Plon, 1900), 72–73.

123. William K. Storey, *Guns, Race, and Power in Colonial South Africa* (Cambridge: Cambridge University Press, 2008), 59, 61–62; Etherington, *Great Treks*, 149–50, 262.

124. Jack Hogan, "'Hardly a Place for a Nervous Old Gentleman to Take a Stroll': Firearms and the Zulu during the Anglo-Zulu War," in *A Cultural History of Firearms in the Age of Empire*, ed. Karen Jones, Giacomo Macola, and David Welch (Farnham: Ashgate, 2013), 145.

125. Read, *Ngoni of Nyasaland*, 46.

126. Ella Kidney, "Native Songs from Nyasaland," *Journal of the Royal African Society* 20, no. 78 (1921): 126, provides no indications of the provenance of this song, but the music sheet Kidney compiled (*Songs of Nyasaland [Central Africa]*, first series [London: Chappell, 1921]) clarifies that it hailed from the Ngoni of Mzimba and that she first heard it in 1911. See also Ruth H. Finnegan, *Oral Literature in Africa* (1970; repr., Cambridge: Open Book, 2012), 204.

127. G. R. Deare, "Eighteen Months with the Last of the Slave Raiders," *Weekend Advertiser* (supplement to the *Natal Advertiser* [Durban, South Africa]), 6 April–11 May 1929. In the typescript copy of the text that I have consulted (Archives of the Livingstone Museum, LM2/4/93/8), the passage quoted above is to be found on pp. 38–39.

128. R. Laws to T. Main, "Livingstonia, Lake Nyassa," 7 February 1879, NLS, MS 7876. For a published version, see *Free Church of Scotland Monthly Record* 203 (June 1879): 136–37.

129. Conner, *Art of the Jere and Maseko Ngoni*, 86.

130. Elmslie, *Among the Wild Ngoni*, 98.

131. Rau, "Mpezeni's Ngoni," 260.

CHAPTER 6: OF "MARTIAL RACES" AND GUNS

1. John McCracken, *A History of Malawi 1859–1966* (Woodbridge, UK: James Currey, 2012), 63.

2. Lewis H. Gann, *A History of Northern Rhodesia: Early Days to 1953* (London: Chatto & Windus, 1964), 89.

3. Thomas Spear, "Neo-traditionalism and the Limits of Invention in British Colonial Africa," *Journal of African History* 44, no. 1 (2003): 3–27.

4. The best accounts are provided by John A. Barnes, *Politics in a Changing Society: A Political History of the Fort Jameson Ngoni* (Cape Town: Oxford University Press, 1954), chap. 3; William E. Rau, "Mpezeni's Ngoni of Eastern Zambia, 1870–1920" (PhD diss., UCLA, 1974), chap. 7; and Harry W. Langworthy, "Introduction: Carl Wiese and Zambezia," in Carl Wiese, *Expedition in East-Central Africa, 1888–1891: A Report*, ed. Harry W. Langworthy (Norman: University of Oklahoma Press, 1983), 3–46.

5. Wiese, *Expedition in East-Central Africa*, appendix D, 368–69.

6. Rau, "Mpezeni's Ngoni," 249.

7. W. Honey to F. C. Warringham, Blantyre, 30 March 1896, and P. W. Forbes to F. C. Warringham, n.p., 7 January 1897, both in Poole Papers. For a description of Fort Partridge, see R. I. Money and S. Kellett Smith, "Explorations in the Country West of Lake Nyasa," *Geographical Journal* 10, no. 2 (1897): 166.

8. Rau, "Mpezeni's Ngoni," 249; Langworthy, "Introduction," 26.

9. A detailed account of the campaign, which saw the deployment of 150 regular soldiers, more than 2,000 gun-bearing irregulars, and one 7-pounder mountain gun, is to be found in "The Mwasi Kazungu Campaign: Extract from Mr. A. J. Swann's Report to Commissioner Johnston," *British Central Africa Gazette* 3, no. 4 (15 February 1896). See also Harry W. Langworthy, "Swahili Influence in the Area between Lake Malawi and the Luangwa River," *African Historical Studies* 4, no. 3 (1971): 601–2.

10. Honey to Warringham, 30 March 1896.

11. Rau, "Mpezeni's Ngoni," 247.

12. Langworthy, "Introduction," 24.

13. R. G. Warton to Secretary (NCEC), Luangeni, 6 August 1896, and P. W. Forbes to Secretary (BSAC), n.p., 9 October 1896, both in Poole Papers; Langworthy, "Introduction," 31–32.

14. Forbes to Secretary, 9 October 1896.

15. G. R. Deare, "Eighteen Months with the Last of the Slave Raiders," *Weekend Advertiser* (supplement to the *Natal Advertiser* [Durban, South Africa]), 6 April–11 May 1929. In the typescript copy of the text that I have consulted (Archives of the Livingstone Museum, LM2/4/93/8), the passage quoted above is to be found on p. 31.

16. I borrow this useful term from Andrew D. Roberts, A *History of Zambia* (London: Heinemann, 1976), 119.

17. First advanced by Barnes, *Politics in a Changing Society*, this reading of intergenerational conflict in Ungoni on the eve of the war of 1898 was accepted and elaborated on by Rau, "Mpezeni's Ngoni," 258–62, and Langworthy, "Introduction," 29.

18. "Local news," *British Central Africa Gazette* 3, no. 22 (1 December 1896), and 4, no. 1 (1 March 1897); "Extracts from Mr. A. J. Swann's Report on the Marimba District for the Year 1896," *British Central Africa Gazette* 4, no. 7 (15 April 1897).

19. Hugo Genthe, "A Trip to Mpeseni's," *British Central Africa Gazette* 4, no. 13 (1 August 1897).

20. Ibid.

21. Ibid.

22. P. W. Forbes to Secretary (BSAC, Salisbury), [Blantyre?], 2 April 1897, Poole Papers.

23. Langworthy, "Introduction," 36.

24. H. H. Johnston to FO, 6 February 1897, in *North Charterland Concession Inquiry: Report to the Governor of Northern Rhodesia by the Commissioner, Mr. Justice Maugham, July 1932* (London: HMSO, 1932), appendix 35.

25. "Mpezeni," in *British Central Africa Gazette* 5, no. 1 (15 January 1898).

26. P. H. Selby to Acting Administrator (Salisbury), [Blantyre?], 16 December 1897, and P. H. Selby to Acting Administrator (Salisbury), [Blantyre?], 31 December 1897, both in Poole Papers.

27. Deare, "Eighteen Months," 41–42.

28. Langworthy, "Introduction," 39.

29. W. H. Manning to Marquess of Salisbury, Zomba, 13 January 1898, encl. in FO to CO, 1 March 1898, NAZ, NW/HC4/2/1, vol. 3.

30. "Mpezeni," *British Central Africa Gazette*.

31. H. E. Brake to A. Sharpe, Luangeni, 20 January 1898, encl. in W. H. Manning to Marquess of Salisbury, Zomba, 17 February 1898, NAUK, FO 2/147. Warringham's police did not take part in the expedition and remained confined in Fort Jameson.

32. Ibid.

33. Ibid.

34. "The Mpezeni Campaign," *British Central Africa Gazette* 5, no. 3 (26 February 1898).

35. Brake to A. Sharpe, Luangeni, 20 January 1898.

36. H. E. Brake to A. Sharpe, Luangeni, 21 January 1898, encl. in Manning to Salisbury, 17 February 1898.

37. Brake to A. Sharpe, Luangeni, 21 January 1898; Rau, "Mpezeni's Ngoni," 271–72.

38. Rau, "Mpezeni's Ngoni," 272.

39. "The Mpezeni Campaign," *British Central Africa Gazette*.

40. Brake to Sharpe, 21 January 1898; Rau, "Mpezeni's Ngoni," 274.

41. W. H. Manning to Marquess of Salisbury, Zomba, 20 May 1898, NAUK, FO2/148.

42. "The Mpezeni Campaign," *British Central Africa Gazette.*

43. "Mpezeni War," *British Central Africa Gazette* 5, no. 2 (5 February 1898); H. L. Daly to Acting Administrator (Salisbury), 4 February 1898, Poole Papers.

44. "The Mpezeni Campaign," *British Central Africa Gazette.* Brake himself admitted that the losses suffered by the Ngoni on the afternoon of 20 January were "unknown"; Brake to Sharpe, 21 January 1898.

45. "The Mpezeni Campaign," *British Central Africa Gazette.*

46. Ibid.

47. See, for instance, Harri Englund's interview with Zilole Zulu, Chief Madzimawe's, 20 April 2012 (in the author's possession). Detailed accounts of the "pacification" of Ungoni between January and February 1898 are to be found in H. E. Brake to A. Sharpe, Chimpinga, 29 January 1898, encl. in Manning to Salisbury, 17 February 1898, and in Manning to Salisbury, 17 February 1898.

48. The quoted description of Ungoni is Warton's; Warton to Secretary (NCEC), 6 August 1896.

49. H. E. Brake to W. H. Manning, Fort Manning, 6 May 1898, encl. in W. H. Manning to Marquess of Salisbury, Zomba, 16 June 1898, NAUK, FO2/148.

50. H. L. Daly to Acting Administrator (Salisbury), [Blantyre?], 17 February 1898, Poole Papers.

51. Ian Linden, "The Maseko Ngoni at Domwe, 1870–1900," in *The Early History of Malawi*, ed. Bridglal Pachai (London: Longman, 1972), 242–43. See also J. L. Nicoll, "Notes on Chikusi's Country," *British Central Africa Gazette* 1, no. 9 (28 June 1894). Gomani is said to have been "16 or 17 years of age" when he succeeded Chikusi in 1891; "Recent Events in Angoniland," *British Central Africa Gazette* 1, no. 16 (14 December 1894).

52. J. L. Nicoll, "Notes on a Journey to Angoniland," *British Central Africa Gazette* 1, no. 3 (20 February 1894).

53. "Raid by Chikusi," *British Central Africa Gazette* 3, no. 19 (15 October 1896).

54. Rev. Chisholm's letter of 24 October 1896, quoted in Melville E. Leslie, "Ncheu in the 1890's," *Society of Malawi Journal* 24, no. 1 (1971): 67–69.

55. The composition and armament of the force are presented differently in "The Angoni War," *British Central Africa Gazette* 3, no. 20 (1 November 1896), and R. C. F. Maugham [Greville], *Africa as I Have Known It: Nyasaland–East Africa–Liberia–Senegal* (London: John Murray, 1929), 166.

56. Maugham, *Africa as I Have Known It*, 167–69.

57. Ibid.; "The Angoni War," *British Central Africa Gazette.*

58. Maugham, *Africa as I Have Known It*, 175–79.

59. "Local News," *British Central Africa Gazette* 3, no. 21 (15 November 1896). Only in April 1898, when a section of the Maseko under Msekandiwana

made a last stand against the British on Domwe Peak, did firearms seem to have been deployed more effectively, forcing a comparatively small CAR party to retire on Dedza until the arrival of reinforcements; J. S. Brogden to F. B. Pearce, Fort Mangoche, 17 May 1898, encl. in W. H. Manning to Marquess of Salisbury, Zomba, 27 May 1898, NAUK, FO 2/148.

60. Daly to Acting Administrator, 17 February 1898.

61. This is not to be confused with the earlier BSAC post among the Chewa of the *Chinunda*. The new Fort Jameson (present-day Chipata) was located in the Ngoni heartland, on the same site as the village of Kapatamoyo, one of Mpezeni's indunas. Rau, "Mpezeni's Ngoni," 294.

62. A. Sharpe to Marquess of Salisbury, Zomba, 8 May 1899, in *North Charterland Concession Inquiry*, appendix 39.

63. The relationship between twentieth-century labor migrancy and pre-colonial notions of honor in southern Africa is touched upon in John Iliffe, *Honour in African History* (Cambridge: Cambridge University Press, 2005), chap. 16.

64. "Report of an interview with Mr. Hayes . . . ," 15 January 1900, encl. in Secretary to J. H. Hayes, Fort Jameson, 18 Jan. 1900, NAZ, BSAC/NER/A3/12/1.

65. Rau, "Mpezeni's Ngoni," 347–50.

66. My reading of Ngoni experiences in colonial armed forces owes much to Iliffe, *Honour in African History*, 234, and Allen F. Isaacman and Barbara S. Isaacman, *Slavery and Beyond: The Making of Men and Chikunda Ethnic Identities in the Unstable World of South-Central Africa, 1750–1920* (Portsmouth, NH: Heinemann, 2004), 296–97.

67. R. G. Warton to P. Forbes, [Beira?], 2 April 1896, in *North Charterland Concession Inquiry*, appendix 28.

68. Warton to Secretary, 6 August 1896.

69. H. H. Johnston, "Memorandum," [London?], 14 July 1896, in *North Charterland Concession Inquiry*, appendix 31.

70. P. W. Forbes, "Memorandum," Reading, 8 March 1898, encl. in H. Canning to FO, London, 14 March 1898, NAZ, NW/HC4/2/1, vol. 3.

71. The figure of 200 is given by Colin Harding, *Far Bugles* (London: Simpkin Marshall, 1933), 82–83; that of 150 in W. H. Manning to A. Sharpe, Zomba, 2 November 1898, NAUK, FO2/149. Alongside a similar number of Tonga recruits, 30 Abercorn (Mbala) men were also sent down to Salisbury at this time. Uncertainty about their fate after the expiration of their contracts precipitated a flurry of correspondence that can be consulted in NAZ, BSAC/NER/A3/11/1. Rau's mistaken claim that Harding was "not able to fill his quota" of Ngoni policemen is due to confusion between the Fort Jameson and Abercorn recruits. Rau, "Mpezeni's Ngoni," 294.

72. C. Harding, "Report of a Journey into Northern Rhodesia," n.d. (late 1898), Archive of the Livingstone Museum, LM 2/3/11/9. I owe this reference to Jack Hogan.

73. Harding, *Far Bugles*, 82–83.

74. Charles van Onselen, "The Role of Collaborators in the Rhodesian Mining Industry, 1900–1935," *African Affairs* 72, no. 289 (1973): 405.

75. The quoted expression is to be found in R. Codrington to Manager (NCEC), "Mpeseni's," 7 September 1898, NAZ, NER/A3/12/1. The cattle losses of 1898 are engraved in Ngoni historical consciousness (see, for instance, group interview, Chief Nzamane's, 22 April 2012). Their actual extent is shrouded in controversy and confusion. By mid-February 1898, the CAR had certainly seized as many as 12,000 head of cattle, roughly 50 percent of the estimated prewar stock (Manning to Salisbury 17 February 1898; F. Young to FO, 4 August 1899, in *North Charterland Concession Inquiry*, appendix 40). What is less clear is how many of these were eventually returned to their original owners. In May of the same year, after admitting that the BSAC had already sold and driven out about 5,000 cattle, Manning proposed that 3,000 be handed back to Mpezeni (Manning to Salisbury, 20 May 1898). A few months later, the same Manning defended the CAR's actions in Ungoni by pointing out that all the confiscated cattle had been "returned to the natives with the exception of about 1,000 handed over as a fine to the British South Africa Co., and about 200 head kept for the use of the troops." (W. H. Manning to Marquess of Salisbury, Zomba, 31 August 1898, NAUK, FO2/148.) On the basis of Manning's figures, then, it seems reasonable to conclude that the Ngoni lost a minimum of 6,200 cattle over the course of 1898 alone.

76. Harding, *Far Bugles*, 82.

77. Timothy J. Stapleton, *African Police and Soldiers in Colonial Zimbabwe, 1923–80* (Rochester, NY: University of Rochester Press, 2011), 22. A nuanced view that points to the significance of both material and cultural forces in shaping the motives of early African *Schutztruppe* in German East Africa is offered by Michelle R. Moyd, *Violent Intermediaries: African Soldiers, Conquest, and Everyday Colonialism in German East Africa* (Athens: Ohio University Press, 2014), esp. chap. 1.

78. Harding, *Far Bugles*, 83.

79. The quoted expression is to be found in David M. Anderson and David Killingray, "Consent, Coercion and Colonial Control: Policing the Empire, 1830–1940," in *Policing the Empire: Government, Authority and Control, 1830–1940*, ed. David M. Anderson and David Killingray (Manchester: Manchester University Press, 1991), 7.

80. W. Manning, Annual Report on the BCAP, encl. in W. H. Manning to Marquess of Salisbury, Zomba, 13 May 1898, NAUK, FO2/147; Risto Marjomaa, "The Martial Spirit: Yao Soldiers in British Service in Nyasaland (Malawi), 1895–1939," *Journal of African History* 44, no. 3 (2003): 419–20; John McCracken, "Coercion and Control in Nyasaland: Aspects of the History of a Colonial Police Force," *Journal of African History* 27, no. 1 (1986): 128.

81. R. Coryndon, "Notes on Police," encl. in R. Coryndon to A. Milner, London, 12 July 1900, NAZ, BSAC/NW/HC4/2/1, vol. 6.

82. C. Monro, "Report on the Barotse Native Police," encl. in H. Marshall Hole to Imperial Secretary (henceforth IS), Kalomo, 8 January 1904, NAZ, BSAC/NW/HC1/2/10.

83. Anderson and Killingray, "Consent, Coercion and Colonial Control," 6.

84. C. Harding to R. Coryndon, [Kazungula], n.d. [but mid-November 1899], encl. in J. Chamberlain to A. Milner, London, 17 March 1900, NAZ, BSAC/NW/HC4/2/1, vol. 6; C. Harding to Secretary (BSAC, Salisbury), Lealui, 15 July 1900, NAZ, BSAC/NW/A6/1/1.

85. F. Hodson, "Annual Return of Military and Naval Resources of Colonies and Protectorates," Kalomo, 13 November 1903, encl. in H. Marshall Hole to A. Lawley, Kalomo, 7 December 1903, NAZ, BASC/NW/HC1/2/10. Though written at the end of 1903, Hodson's return describes the situation obtaining in 1902.

86. C. Harding to R. T. Coryndon, Kalomo, 17 February 1903, encl. in R. T. Coryndon to Secretary (BSAC), Kalomo, 6 March 1903, NAZ, BSAC/NW/A2/2/3.

87. R. T. Coryndon to A. Milner, Kalomo, 25 May 1903, encl. in Secretary to the Administrator to Secretary (BSAC), Kalomo, 25 May 1903, NAZ, BSAC/NW/A2/2/3; William V. Brelsford, *The Story of the Northern Rhodesia Regiment* (1954; repr., Bromley, UK: Galago, 1990), 20–21; Tim B. Wright, *The History of the Northern Rhodesia Police* (Bristol: British Empire & Commonwealth Museum Press, 2001), 47.

88. C. Harding to IS, Kalomo, 11 January 1905, encl. in Secretary to the Administrator (NWR) to IS, Kalomo, 14 January 1905, NAZ, BSAC/NW/HC1/2/15.

89. C. Harding to IS, Kalomo, 23 November 1904, encl. in R. T. Coryndon to IS, Kalomo, 24 January 1904, NAZ, BSAC/NW/HC1/2/14.

90. "Nominal roll of natives enlisted for service in Barotse Native Police," Fort Jameson, 4 October 1904, encl. in Harding to IS, 11 January 1905.

91. "Annual Return of Military and Naval Resources of Colonies and Protectorates," encl. in C. Harding to IS, [Kalomo], 18 January 1905, NAZ, BSAC/NW/HC1/2/15.

92. [F. Hodson], Untitled memorandum on the BNP, Kalomo, 9 October 1905, encl. in F. Hodson to IS, [Kalomo], 9 October 1905, NAZ, BSAC/NW/HC1/2/22.

93. "Annual Report on Barotse Native Police for Year 1908–1909," NAZ, BSAC/NR/A5/1/2.

94. In 1909, North-Eastern Rhodesia's armed police consisted of 310 Africans; of these, 109 were classified as Bemba and 71 as Ngoni. Acting Adminstrator (NER), "Annual Return of Military and Naval Resources of North-Eastern Rhodesia," 31 December 1909, encl. in Secretary to the Administrator (NER) to IS, [Fort Jameson?], 2 August 1910, NAZ, BSAC/NER/A2/3/1, vol. 3.

95. Heather Streets, *Martial Races: The Military, Race and Masculinity in British Imperial Culture, 1857–1914* (Manchester: Manchester University Press, 2004), 3, 94.

96. Timothy H. Parsons, *The African Rank-and-File: Social Implications of Colonial Military Service in the King's African Rifles, 1902–1964* (Portsmouth, NH: Heinemann, 1999), 9. See also, by the same author, "'Wakamba Warriors Are Soldiers of the Queen': The Evolution of the Kamba as a Martial Race, 1890–1970," *Ethnohistory* 46, no. 4 (1999): 671–701.

97. Edward I. Steinhart, *Black Poachers, White Hunters: A Social History of Hunting in Colonial Kenya* (Oxford: James Currey 2006), 57.

98. Myles Osborne, "Controlling Development: 'Martial Race' and Empire in Kenya, 1945–59," *Journal of Imperial and Commonwealth History* 42, no. 3 (2014): 471.

99. Parsons, "'Wakamba Warriors,'" 673.

100. Moyd, *Violent Intermediaries*, 113–14. See also 62–63 for the short-lived Mozambican "Zulu" (i.e., Shangaan) presence in the German colonial army.

101. A. Cree, "Regimental miscellanea," in Brelsford, *Story of the Northern Rhodesia Regiment*, 117; Hodson, "Annual Return."

102. See, e.g., Monro, "Report on the Barotse Native Police"; Harding to IS, 23 November 1904; "Report on Barotse Native Police, 1910–1911," NAZ, BSAC/NR/A5/1/2.

103. The award of "Marksman Badges" is mentioned repeatedly in regimental orders dating to 1907–1909, NAZ, BSAC/NW/B2/1/6.

104. Michelle Moyd has convincingly demonstrated that the *askari's* ambition to "become big men" was significantly fostered by their "ability to coerce and expropriate resources" from the peoples they conquered and/or policed in German East Africa. Moyd, *Violent Intermediaries*, 18–19, 123, 133.

105. H. Marshall Hole to J. Carden, Kalomo, 19 August 1903, encl. in H. Marshall Hole to A. Lawley, Kalomo, 19 August 1903, NAZ, BSAC/NW/HC1/2/9.

106. Quoted in J. Carden to H. Marshall Hole, Linungwe, 30 August 1903, encl. in H. Marshall Hole to A. Lawley, Kalomo, 21 September 1903, NAZ, BSAC/NW/HC1/2/10.

107. Monro, "Report on the Barotse Native Police."

108. Regimental orders, 19 September 1907, 16 May and 26 September 1908, NAZ, BSAC/NW/B2/1/6.

109. Regimental orders, 19 September and 31 December 1907, 15 and 20 April 1908, NAZ, BSAC/NW/B2/1/6.

110. Regimental orders, 25 January, 25 February, and 7 April 1908, NAZ, BSAC/NW/B2/1/6.

CONCLUSION

1. Among the dozens of reports and similar publications that could be cited, see, for instance, International Crisis Group, "Disarmament in the Congo: Investing in Conflict Prevention," ICG Africa Briefing Paper 4, 12 June 2001, http://www.crisisgroup.org/~/media/Files/africa/central-africa/drcongo/Disarmament%20in%20the%20Congo%20Investing%20in%20Conflict%20Prevention; International Crisis Group, "Congo: A Comprehensive Strategy to Disarm the FDLR," ICG Africa Report 151, 9 July 2009, http://www.crisisgroup.org/~/media/Files/africa/central-africa/dr-congo/151%20Congo%20-%20A%20Comprehensive%20Strategy%20to%20Disarm%20the%20FDLR%20-%20ENGLISH.pdf; Global Witness, "'Faced with a Gun, What Can You Do?' War and the Militarisation of Mining in Eastern Congo," July 2009, http://www.globalwitness.org/sites/default/files/pdfs/report_en_final_0.pdf.

2. Koen Vlassenroot, "A Societal View on Violence and War: Conflict and Militia Formation in Eastern Congo," in *Violence, Political Culture and Development in Africa*, ed. Preben Kaarsholm (Oxford: James Currey, 2006), 51, 57, 59; Séverine Autesserre, *The Trouble with the Congo: Local Violence and the Failure of International Peacebuilding* (Cambridge: Cambridge University Press, 2010).

3. On Sierra Leone's Revolutionary United Front and the Kamajo militias pitted against it in the 1990s, see, for instance, Krijn Peters and Paul Richards, "'Why We Fight': Voices of Youth Combatants in Sierra Leone," *Africa* 68, no. 2 (1998): 183–210.

4. Christopher A. Bayly, *The Birth of the Modern World, 1780–1914: Global Connections and Comparisons* (Oxford: Blackwell, 2004), 12.

5. Luca Jourdan, "Mayi-Mayi: Young Rebels in Kivu, DRC," *Africa Development* 36, nos. 3–4 (2011): 103.

6. Jean-Claude Willame, quoted in Autesserre, *Trouble with the Congo*, 148. See also Luca Jourdan, *Generazione Kalashnikov: Un antropologo dentro la guerra in Congo* (Rome: Laterza, 2010).

7. In Koli Jean Bofane, *Congo Inc.: Le testament de Bismarck* (Arles: Actes Sud, 2014), 122–23.

8. Henry M. Stanley, *Through the Dark Continent*, 2 vols. (London: Sampson Low, Marston, Seable & Rivington, 1878), 2:129–30. On the *waungwana* of Manyema, see Melvin E. Page, "The Manyema Hordes of Tippu Tip: A Case Study in Social Stratification and the Slave Trade in Eastern Africa," *International Journal of African Historical Studies* 7, no. 1 (1974): 69–84.

9. Verney L. Cameron, *Across Africa* (New York: Harper & Brothers, 1877), 259.

10. Hermann von Wissmann, *My Second Journey through Equatorial Africa: From the Congo to the Zambesi, in the Years 1886 and 1887* (London: Chatto & Windus, 1891), 188–91.

11. Richard J. Reid, *War in Pre-Colonial Eastern Africa: The Patterns and Meanings of State-Level Conflict in the Nineteenth Century* (Oxford: James Currey, 2007), 119. On *ruga ruga* in northern Katanga, see also Allen F. Roberts, *A Dance of Assassins: Performing Early Colonial Hegemony in the Congo* (Bloomington: Indiana University Press, 2013), 21–22, 71–72.

12. Kennedy A. Mkutu, *Guns and Governance in the Rift Valley: Pastoralist Conflict and Small Arms* (Oxford: James Currey, 2008), 44; Ben Knighton, "The State as Raider among the Karamojong: 'Where There Are No Guns, They Use the Threat of Guns,'" *Africa: Journal of the International African Institute* 73, no. 3 (2003): esp. 432–23, 436–38.

13. Sharon E. Hutchinson, *Nuer Dilemmas: Coping with Money, War, and the State* (Berkeley: University of California Press, 1996), 150.

14. Important trend-setters, in this regard, were K. B. Wilson, "Cults of Violence and Counter-Violence in Mozambique," *Journal of Southern African Studies* 18, no. 3 (1992): 527–82, and Stephen Ellis, *The Mask of Anarchy:*

The Destruction of Liberia and the Religious Dimension of an African Civil War (London: Hurst, 1999).

15. Wilson, "Cults of Violence," 565–56.

16. Richard J. Reid, "Past and Presentism: The 'Precolonial' and the Fore-shortening of African History," *Journal of African History* 52, no. 2 (2011): 153.

17. Ibid., 136.

Bibliography

PRIMARY SOURCES

Archival Deposits

Full archival citations are given in the notes. Listed below are re-
positories, and the series and collections, upon which I have relied
most heavily.

ZAMBIA

National Archives of Zambia (NAZ), Lusaka

Select files belonging to the following series: BSAC/NW (Records
of the British South Africa Company Administration, North-Western
Rhodesia); BSAC/NER (Records of the British South Africa Company
Administration, North-Eastern Rhodesia); BSAC/NR (Records of the
British South Africa Company Administration, Northern Rhodesia); ZA
(Secretary for Native Affairs); and KDE (Barotse), KSE (Mwinilunga),
and KDD (Kasempa)

Archives of the Livingstone Museum, Livingstone

E. H. Lane Poole's Papers

Miscellaneous material by Robert Coryndon, G. R. Deare, and Colin
Harding

UNITED KINGDOM

National Archives of the United Kingdom (NAUK), Kew, London

Select files from: FO 84 (Slave Trade Department and Successors);
FO 2 (General Correspondence before 1906, Africa); FO 10 (General
Correspondence before 1906, Belgium)

Archives of the London School of Economics, London

Margaret Read's Papers

National Library of Scotland (NLS), Edinburgh

Livingstonia Mission's Papers

Centre for Research Collections, Edinburgh University Library, Edinburgh

Papers of Robert Laws

BELGIUM

Musée royal de l'Afrique centrale (MRAC), Tervuren

Clément Brasseur's Papers (BP)

Léon Guebels's Papers (GP)

Archives africaines de l'ex-ministre des Affaires africaines (AA), Ministère des Affaires étrangères, Brussels

Select files belonging to the AE (Archives du Département des Affaires étrangères de l'État Independent du Congo) and the AIMO (Fonds Affaires indigènes main-d'œuvre) series.

Bibliothèque africaine at the Bibliothèque des Affaires étrangères, Brussels

"Documents du Zaïre Colonial—Collection J. L. Vellut"

PORTUGAL

Biblioteca pública municipal do Porto (BPMP), Porto

Antonio Francisco da Silva Porto, "Viagens e apontamentos de um portuense em África," vols. 3 to 5, August 1862–April 1869

Official Publications

British Central Africa Gazette (Zomba)
Bulletin Officiel du Congo Belge (Brussels)
Cape of Good Hope Government Gazette (Cape Town)
Codes et lois du Congo Belge. Comp. Octave Louwers and Iwan Grenade, with C. Kuck. Brussels: Weissenbruch, 1927.
Lois en vigueur dans l'État Indépendant du Congo. Comp. Octave Louwers. Brussels: Weissenbruch, 1905.
North Charterland Concession Inquiry: Report to the Governor of Northern Rhodesia by the Commissioner, Mr. Justice Maugham, July 1932. London: HMSO, 1932.
North-Eastern Rhodesia Government Gazette (Fort Jameson [Chipata])
Northern Rhodesia Government Gazette (Livingstone/Lusaka)

MISSIONARY PERIODICALS

Echoes of Service: A Record of Labour in the Lord's Name (Bath, UK; Plymouth Brethren/Christian Missions in Many Lands)
Free Church of Scotland Monthly Record (Edinburgh)
South Africa General Mission Pioneer (London; called *The South African Pioneer* until 1919)

OTHER PUBLISHED PRIMARY SOURCES

Arnot, Frederick S. *Bihé and Garenganze; or, Four Years' Further Work and Travel in Central Africa.* London: James E. Hawkins, 1893.
———. *Garenganze; or, Seven Years' Pioneer Mission Work in Central Africa.* London: James E. Hawkins, 1889.
Baptista, Pedro João. "Journey of the 'Pombeiros' from Angola to the Rios de Senna." In *The Lands of Cazembe*, edited by Richard F. Burton, 165–244. London: John Murray, 1873.
Brasseur, Clément. "Lettre sur le Katanga." *La Belgique Coloniale* 2, no. 17 (1896): 197–200.
———. "Lettre sur le Katanga." *La Belgique Coloniale* 3, no. 20 (1897): 233–34.
Briart, Paul. *Aux sources du fleuve Congo: Carnets du Katanga (1890–1893).* Edited by Dominique Ryelandt. Paris: L'Harmattan, 2003.
Bruwer, J. P. *Mbiri ya Angoni.* Mkhoma, Malawi: DRC Mission Press, 1941.
Buchanan, John. *The Shirè Highlands (East Central Africa) as Colony and Mission.* Edinburgh: William Blackwood, 1885.
Cameron, Verney L. *Across Africa.* New York: Harper & Brothers, 1877.
Campbell, Dugald. *In the Heart of Bantuland.* London: Seeley, Service, 1922.
———. Letter to H.R. Fox Bourne, Johnston Falls, 14 May 1904, *The Aborigines' Friend* (October 1904): 201–14.
Capello, Hermenegildo Carlos de Brito, and Roberto Ivens. *De Angola á contra-costa: Descripção de uma viagem atravez do continente africano.* 2 vols. Lisbon: Imprensa Nacional, 1886.
Carvalho, Henrique A. Dias de. *Expedição portugueza ao Muatiânvua: Ethnographia e historia tradicional dos povos da Lunda.* Lisbon: Imprensa Nacional, 1890.
Chapman, James. *Travels in the Interior of South Africa, 1849–1863.* Edited by E. C. Tabler. 2 vols. 1868. Reprint, Cape Town: A. A. Balkema, 1971.
Chibambo, Yesaya M. *My Ngoni of Nyasaland.* Translated by Charles Stuart. London: United Society for Christian Literature, n.d. [1942].
Chibanza, Simon Jilundu. *Kaonde History.* Part 2 of J. T. Munday and Simon Jilundu Chibanza, *Central Bantu Historical Texts*, vol.1. Lusaka: Rhodes-Livingstone Institute, 1961.
Coillard, François. *On the Threshold of Central Africa: A Record of Twenty Years' Pioneering among the Barotse of the Upper Zambesi.* 1897. Reprint, London: Frank Cass, 1971.
Cotterill, H. B. "On the Nyassa and a Journey from the North End to Zanzibar." *Proceedings of the Royal Geographical Society* 22, no. 4 (1877–78): 233–51.

Delcommune, Alexandre. *Vingt années de vie africaine: Récits de voyages, d'aventures et d'exploration au Congo belge, 1874–1893.* 2 vols. Brussels: Ferdinand Larcier, 1922.

Elmslie, W. A. *Among the Wild Ngoni.* Edinburgh: Oliphant, Anderson & Ferrier, 1899.

Foà, Édouard. *Du Cap au Lac Nyassa.* 1897. Reprint, Paris: Librairie Plon, 1901.

———. *Résultats scientifiques des voyages en Afrique.* Paris: Imprimerie Nationale, 1908.

———. *La traversée de l'Afrique du Zambèze au Congo français.* Paris: Librairie Plon, 1900.

Fraser, Donald. *Winning a Primitive People: Sixteen Years' Work among the Warlike Tribe of the Ngoni and the Senga and Tumbuka Peoples of Central Africa.* 1914. Reprint, London: Seeley, Service, 1922.

Gamitto, Antonio C. P. *King Kazembe and the Marave, Cheva, Bisa, Bemba, Lunda and Other Peoples of Southern Africa.* Translated by Ian G. Cunnison. 2 vols. Lisbon: Junta de Investigações do Ultramar, 1960.

Gelfand, Michael, ed. *Gubulawayo and Beyond: Letters and Journals of the Early Jesuit Missionaries to Zambesia (1879–1887).* London: Geoffrey Chapman, 1968.

Gibbons, Alfred St. H. *Africa from South to North through Marotseland.* 2 vols. London: John Lane, 1904.

Giraud, Victor. *Les lacs de l'Afrique équatoriale.* Paris: Librairie Hachette, 1890.

Habib, Said ibn. "Narrative of Said bin Habeeb, an Arab Inhabitant of Zanzibar." *Transactions of the Bombay Geographical Society* 15 (1860): 146–48.

Harding, Colin. *Far Bugles.* London: Simpkin Marshall, 1933.

———. *In Remotest Barotseland.* London: Hurst and Blackett, 1904.

Holub, Emil. *Emil Holub's Travels North of the Zambezi, 1885–6.* Edited by Ladislav Holy. Translated by Christa Johns. Manchester: Manchester University Press, 1975.

———. *Seven Years in South Africa: Travels, Researches, and Hunting Adventures between the Diamond-Fields and the Zambesi (1872–79).* Translated by Ellen E. Frewer. 2 vols. London: Sampson Low, Marston, Searle & Rivington, 1881.

Horton, A. E., comp. *A Dictionary of Luvale.* El Monte, CA: Rahn Brothers Printing and Lithographing, 1953.

Jalla, Adolphe D., comp. *English-Sikololo Dictionary.* Torre Pellice, Italy: Imprimerie Alpine, 1917.

———. *Litaba za Sichaba sa ma-Lozi.* 1910. Reprint, Dundee, South Africa: Ebenezer Press, 1922.

———. *Pionniers parmi les ma-Rotse.* Florence: Imprimerie Claudienne, 1903.

Johnson, William P. *My African Reminiscences, 1875–1895.* London: Universities' Mission to Central Africa, n.d. [1924?].

Kerr, Walter M. *The Far Interior: A Narrative of Travel and Adventure from the Cape of Good Hope across the Zambesi to the Lake Regions of Central Africa.* 2 vols. Boston: Houghton, Mifflin, 1886.

Kidney, Ella. "Native Songs from Nyasaland." *Journal of the Royal African Society* 20, no. 78 (1921): 116–26.

———. *Songs of Nyasaland (Central Africa)*. First series. London: Chappell, 1921.

Last, J. T. "A Journey from Blantyre to Angoni-Land and Back." *Proceedings of the Royal Geographical Society* 9, no. 3 (1887): 177–87.

Laws, Robert. "Journey along Part of the Western Side of Lake Nyassa, in 1878." *Proceedings of the Royal Geographical Society* 1, no. 5 (1879): 305–24.

Le Marinel, Paul. *Carnets de route dans l'État indépendant du Congo de 1887 à 1910*. Brussels: Éditions Progress, 1991.

Linden, Ian, ed. and trans. "Mponda Mission Diary, 1889–1891." *International Journal of African Historical Studies* 7, nos. 2–4 (1974): 272–303, 493–515, 688–728; 8, no. 1 (1975): 111–35.

———. "Some Oral Traditions from the Maseko Ngoni." *Society of Malawi Journal* 24, no. 2 (1971): 60–73.

Livingstone, David. *The Last Journals of David Livingstone, in Central Africa, from 1865 to His Death*. Edited by Horace Waller. 2 vols. London: John Murray, 1874.

———. *Livingstone's African Journal, 1853–1856*. Edited by Isaac Schapera. 2 vols. London: Chatto & Windus, 1963.

———. *Livingstone's Private Journals, 1851–1853*. Edited by Isaac Schapera. London: Chatto & Windus, 1960.

———. *Missionary Travels and Researches in South Africa*. New York: Harper & Brothers, 1858.

Livingstone, David, and Charles Livingstone. *Narrative of an Expedition to the Zambezi and Its Tributaries*. London: John Murray, 1865.

Livingstone, W. P. *Laws of Livingstonia: A Narrative of Missionary Adventure and Achievement*. London: Hodder and Stoughton, 1921.

Lugard, Frederick D. *The Rise of Our East African Empire*. 2 vols. Edinburgh: William Blackwood, 1893.

Magyar, László [Ladislaus]. "Ladislaus Magyar's Erforschung von Inner-Afrika." *Petermann's geographische Mitteilungen* 6 (1860): 227–37.

Maugham [Greville], R. C. F. *Africa as I Have Known It: Nyasaland–East Africa–Liberia–Senegal*. London: John Murray, 1929.

Melland, Frank H. *In Witch-Bound Africa: An Account of the Primitive Kaonde Tribe and Their Beliefs*. 1923. Reprint, London: Frank Cass, 1967.

Moir, Frederick L. M. *After Livingstone: An African Trade Romance*. London: Hodder and Stoughton, n.d. [1923].

Moloney, Joseph A. *With Captain Stairs to Katanga*. 1893. Reprint, London: Jeppestown Press, 2007.

Money, R. I., and S. Kellett Smith. "Explorations in the Country West of Lake Nyasa," *Geographical Journal* 10, no. 2 (1897): 146–72.

Morel, Edmund D. *King Leopold's Rule in Africa*. London: William Heinemann, 1904.

Mukanda Bantu. "Les mémoires de Mukande Bantu," *Bulletin de la Société Belge d'Études Coloniales* 27, nos. 5–6 (1919): 251–77, nos. 9–10 (1919): 497–521.

Mwata Kazembe XIV [Shadreck Chinyanta Nankula]. *Central Bantu Histori-cal Texts*, vol. 2, *Historical Traditions of the Eastern Lunda*. Translated by Ian G. Cunnison. Lusaka: Rhodes-Livingstone Institute, 1962.

Mwenda Munongo, Antoine, ed. and trans. "Chants historiques des Bayeke: Recueillis à Bunkeya et ailleurs." *Problèmes Sociaux Congolais* 77 (1967): 35–139.

Pirouet, H. G. "'The Gates of Hell Shall Not Prevail.'" *South Africa General Mission Pioneer* 36, no. 4 (April 1923): 42–44.

Read, Margaret. "Songs of the Ngoni People." *Bantu Studies* 11, no. 1 (1937): 1–35.

Reichard, Paul. "Herr Paul Reichard: Bericht über seine Reisen in Ostafrika und dem Quellgebiet des Kongo." *Verhandlungen der Gesellschaft für Erdkunde zu Berlin* 13, no. 2 (1886): 107–25.

———. "Bericht von Paul Reichard über die Reise nach Urua und Katanga." *Mittheilungen der Afrikanischen Gesellschaft in Deutschland* 4, no. 5 (1885): 303–9.

———. "Le Katanga." *Congo* (Brussels) 1, no. 3 (1930): 473–79.

Sambeek, Jan van, et al. *Ifya Bukaya: Third Bemba Reader*. Chilubula, Zam-bia: White Fathers, 1932.

———. *Ifya Bukaya: Fourth Bemba Reader*. Chilubula, Zambia: White Fa-thers, n.d.

Sangambo, Mose K. *The History of the Luvale People and Their Chieftainship*. 2nd ed. Zambezi, Zambia: Mize Palace, n.d. [1985?].

Sebestyen, Eva, and Jan Vansina, ed. and trans. "Angola's Eastern Hinterland in the 1750s: A Text Edition and Translation of Manoel Correia Leitão's 'Voyage' (1755–1756)." *History in Africa* 26 (1999): 299–364.

Selous, Frederick Courteney. *Travel and Adventure in South-East Africa*. Lon-don: Rowland Ward, 1893.

Serpa Pinto, Alexandre A. da Rocha de. *How I Crossed Africa*. Translated by Alfred Elwes. 2 vols. London: Sampson Low, Marston, Searle, & Riving-ton, 1881.

Sharpe, Alfred. "Alfred Sharpe's Travels in the Northern Province and Ka-tanga." *Northern Rhodesia Journal* 3, no. 3 (1957): 210–19.

———. "A Journey through the Country Lying between the Shire and Loangwa Rivers," *Proceedings of the Royal Geographical Society* 12, no. 3 (1890): 150–57.

Silva Porto, Antonio Francisco da. *Viagens e apontamentos de um portuense em África*. Lisbon: Agência Geral das Colónias, 1942.

———. *Viagens e apontamentos de um portuense em África*. Edited by Maria Emilia Madeira Santos. Coimbra, Portugal: Biblioteca Geral da Univer-sidade de Coimbra, 1986.

Silva Teixeira, Alexandre da (?). "Derrota de Benguella para o sertão." In *An-gola: Apontamentos sôbre a colonização dos planaltos e litoral do sul de Angola*, edited by Alfredo de Albuquerque Felner, 2:13–27. 3 vols. Lisbon: Agência Geral das Colónias, 1940.

Silva Teixeira, Alexandre da. "Relação da viagem que fiz deste cidade de Benguella para as terras de Lovar no anno de mil setecentos noventa e quatro." *Arquivos de Angola* (Luanda) 1, no. 4 (1935): doc. X.

Smith, Edwin W., and Andrew M. Dale. *The Ila-Speaking Peoples of Northern Rhodesia*. 2 vols. London: Macmillan, 1920.

Stairs, William G. "De Zanzibar au Katanga: Journal du Capitaine Stairs (1890–1891)." *Le Congo Illustré* 2, nos. 21–26 (1893): 166–67, 173–75, 181–83, 189–91, 197–99, 205–7.

Stanley, Henry M. *Through the Dark Continent*. 2 vols. London: Sampson Low, Marston, Seable & Rivington, 1878.

Stevenson-Hamilton, James. *The Barotseland Journal of James Stevenson-Hamilton, 1898–1899*. Edited by J. P. R. Wallis. London: Chatto & Windus, 1953.

Stewart, James. "The Second Circumnavigation of Lake Nyassa." *Proceedings of the Royal Geographical Society* 1, no. 5 (1879): 289–304.

Tippu Tip. *Maisha ya Hamed bin Muhammed el Murjebi, yaani Tippu Tip, kwa maneno yake mwenyewe*. Edited and translated by Wilfred H. Whiteley. 1958–59. Reprint, Nairobi: East African Literature Bureau, 1974.

Venning, J. H. "Early Days in Balovale." *Northern Rhodesia Journal* 2, no. 6 (1955): 53–57.

Verbeek, Léon. *Le chasseur africain et son monde: Chansons de chasse du sud-est du Katanga*. Tervuren, Belgium: Musée royal de l'Afrique centrale, 2007. http://www.africamuseum.be/research/publications/rmca/online/chansons%20de%20chasse.pdf.

———. *L'histoire dans les chants et les danses populaires: La zone culturelle Bemba du Haut-Shaba (Zaïre)*. Enquêtes et documents d'histoire Africaine 10. Louvain-la-Neuve, Belgium: Centre d'histoire de l'Afrique, 1992.

Verdick, Edgard. *Les premiers jours au Katanga (1890–1903)*. Brussels: Comité spécial du Katanga, 1952.

Westbeech, George. "The Diary of George Westbeech: 1885–1888." In *Trade and Travel in Early Barotseland*, edited by E. C. Tabler, 23–102. London: Chatto & Windus, 1963.

Wiese, Carl. "Beiträge zur geschichte der Zulu im norden des Zambesi, namentlich der Angoni." *Zeitschrift für Ethnologie* 32 (1900): 181–201.

———. *Expedition in East-Central Africa, 1888–1891: A Report*. Edited by Harry W. Langworthy. Translated by Donald Ramos. Norman: University of Oklahoma Press, 1983

Wissmann, Hermann von. *My Second Journey through Equatorial Africa: From the Congo to the Zambesi, in the Years 1886 and 1887*. London: Chatto & Windus, 1891.

Woods, R. E. Broughall, comp. *A Short Introductory Dictionary of the Kaonde Language with English-Kaonde Appendix*. London: Religious Tract Society, 1924.

Young, Edward D. *Nyassa: A Journal of Adventures whilst Exploring Lake Nyassa, Central Africa, and Establishing the Settlement of "Livingstonia."* London: John Murray, 1877.

Young, T. Cullen. *Notes on the History of the Tumbuka-Kamanga Peoples in the Northern Province of Nyasaland*. 1932. Reprint, London: Frank Cass, 1970.

Interviews and Personal Communications

Birmingham, David. Email to author, 26 March 2011.
Dubrunfaut, Paul. Personal communication to author, Brussels, 30 June 2012.
Group interview. Author's interview, Chief Nzamane's, Zambia, 22 April 2012.
Hoover, J. Jeffrey. Email to author, 4 May 2013.
Jaeger, Dirk. Email to author, 16 September 2013.
Kidyamba, J. Mwidye. Author's interview, Bunkeya, DR Congo, 5 August 2011.
Mwenda Numbi, B. Author's interview, Bunkeya, DR Congo, 7 August 2011.
Nshita Kafuku, Mwami Mukonki VIII. Author's interview, Bunkeya, DR Congo, 8 August 2011.
Zilole Zulu. Interview by Harri Englund, Chief Madzimawe's, Zambia, 20 April 2012.

SECONDARY SOURCES

Books and Articles

Akyeampong, Emmanuel. *Drink, Power, and Cultural Change: A Social History of Alcohol in Ghana, c. 1800 to Recent Times*. Portsmouth, NH: Heinemann, 1996.
Allman, Jean. "Fashioning Africa: Power and the Politics of Dress." In *Fashioning Africa: Power and the Politics of Dress*, edited by Jean Allman, 1–10. Bloomington: Indiana University Press, 2004.
Alpers, Edward A. *Ivory and Slaves in East Central Africa: Changing Patterns of International Trade to the Later Nineteenth Century*. London: Heinemann, 1975.
———. "The Yao in Malawi: The Importance of Local Research." In *The Early History of Malawi*, edited by Bridglal Pachai, 168–78. London: Longman, 1972.
Anderson, David M., and David Killingray. "Consent, Coercion and Colonial Control: Policing the Empire, 1830–1940." In *Policing the Empire: Government, Authority and Control, 1830–1940*, edited by David M. Anderson and David Killingray, 1–15. Manchester: Manchester University Press, 1991.
Appadurai, Arjun, ed. *The Social Life of Things: Commodities in Cultural Perspective*. Cambridge: Cambridge University Press, 1986.
Arnold, David. "Europe, Technology, and Colonialism in the 20th Century." *History and Technology* 21, no. 1 (2005): 85–106.
Arnold, David, and Erich DeWald. "Cycles of Empowerment? The Bicycle and Everyday Technology in Colonial India and Vietnam." *Comparative Studies in Society and History* 53, no. 4 (2011): 971–96.
Atkinson, Ronald R. *The Roots of Ethnicity: The Origins of the Acholi of Uganda before 1800*. Philadelphia: University of Pennsylvania Press, 1994.

Atmore, Anthony, J. M. Chirenje, and S. I. Mudenge. "Firearms in South Central Africa." *Journal of African History* 12, no. 4 (1971): 545–56.

Auslander, Leora, et al. "AHR Conversation: Historians and the Study of Material Culture." *American Historical Review* 114, no. 5 (2009): 1355–404.

Austen, Ralph A., and Daniel R. Headrick. "The Role of Technology in the African Past." *African Studies Review* 26, nos. 3–4 (1983): 163–84.

Autesserre, Séverine. *The Trouble with the Congo: Local Violence and the Failure of International Peacebuilding.* Cambridge: Cambridge University Press, 2010.

Balzac, Honoré de. *The Human Comedy: Selected Stories.* New York: New York Review Books, 2014.

Bantje, Han. *Kaonde Song and Ritual.* Annales du musée royal de l'Afrique centrale 95. Tervuren, Belgium: Musée royal de l'Afrique centrale, 1978.

Barnes, John A. *Politics in a Changing Society: A Political History of the Fort Jameson Ngoni.* Cape Town: Oxford University Press, 1954.

Bayly, Christopher A. *The Birth of the Modern World, 1780–1914: Global Connections and Comparisons.* Oxford: Blackwell, 2004.

Berg, Gerald M. "The Sacred Musket: Tactics, Technology, and Power in Eighteenth-Century Madagascar." *Comparative Studies in Society and History* 27, no. 2 (1985): 261–79.

Bersselaar, Dmitri van den. *The King of Drinks: Schnapps Gin from Modernity to Tradition.* Leiden: Brill, 2007.

Bijker, Wiebe E., and John Law. "General Introduction." In *Shaping Technology/Building Society: Studies in Sociotechnical Change,* edited by Wiebe E. Bijker and John Law, 1–16. Cambridge, MA: MIT Press, 1992.

Bijker, Wiebe E., Thomas P. Hughes, and Trevor J. Pinch, eds. *The Social Construction of Technological Systems: New Directions in the Sociology and History of Technology.* Cambridge, MA: MIT Press, 1987.

Binsbergen, Wim van. *Tears of Rain: Ethnicity and History in Central Western Zambia.* London: Kegan Paul International, 1992.

Birmingham, David. *Central Africa to 1870: Zambezi, Zaïre and the South Atlantic.* Cambridge: Cambridge University Press, 1981.

Bofane, In Koli Jean. *Congo Inc.: Le testament de Bismarck.* Arles: Actes Sud, 2014.

Boivin, Nicole. *Material Cultures, Material Minds: The Impact of Things on Human Thought, Society, and Evolution.* Cambridge: Cambridge University Press, 2010.

Bonello, Julie. "The Development of Early Settler Identity in Southern Rhodesia: 1890–1914." *International Journal of African Historical Studies* 43, no. 2 (2010): 341–67.

Bontinck, François. "Derrota de Benguella para o sertão: critique d'authenticité." *Bulletin des Séances de l'Académie Royale des Sciences d'Outre-Mer,* n.s., 23, no. 3 (1977): 279–300.

———. "La double traversée de l'Afrique par trois 'Arabes' de Zanzibar (1845–1860)." *Études d'Histoire Africaine* 6 (1974): 5–53.

Brelsford, William V. *The Story of the Northern Rhodesia Regiment*. 1954. Reprint, Bromley, UK: Galago, 1990.

Bridges, Roy C. "Explorers' Texts and the Problem of Reactions by Non-Literate Peoples: Some Nineteenth-Century East African Examples." *Studies in Travel Writing* 2, no. 1 (1998): 65–84.

———. "Nineteenth-Century East African Travel Records." In "European Sources for Sub-Saharan Africa before 1900: Use and Abuse," edited by Beatrix Heintze and Adam Jones, special issue, *Paideuma* 33 (1987): 179–96.

Burke, Timothy. *Lifebuoy Men, Lux Women: Commodification, Consumption, and Cleanliness in Modern Zimbabwe*. Durham: Duke University Press, 1996.

Candido, Mariana P. *An African Slaving Port and the Atlantic World: Benguela and Its Hinterland*. Cambridge: Cambridge University Press, 2013.

———. "Merchants and the Business of the Slave Trade at Benguela, 1750–1850." *African Economic History* 35, no. 1 (2007): 1–30.

Caplan, Gerald L. *The Elites of Barotseland, 1878–1969: A Political History of Zambia's Western Province*. London: C. Hurst, 1970.

Chew, Emrys. *Arming the Periphery: The Arms Trade in the Indian Ocean during the Age of Global Empire*. New York: Palgrave Macmillan, 2012.

Clarence-Smith, William G. "A Note on the 'Ecole des *Annales*' and the Historiography of Africa." *History in Africa* 4 (1977): 275–81.

Cohen, Deborah. *Household Gods: The British and Their Possessions*. New Haven: Yale University Press, 2006.

Comaroff, Jean. "The Empire's Old Clothes: Fashioning the Colonial Subject." In *Cross-Cultural Consumption: Global Markets, Local Realities*, edited by David Howes, 19–38. London: Routledge, 1996.

Comaroff, John L., and Jean Comaroff. *Of Revelation and Revolution*. Vol. 2, *The Dialectics of Modernity on a South African Frontier*. Chicago: University of Chicago Press, 1997.

Conner, Michael W. *The Art of the Jere and Maseko Ngoni of Malawi, 1818–1964*. New York: Man's Heritage Press, 1993.

Crehan, Kate. *The Fractured Community: Landscapes of Power and Gender in Rural Zambia*. Berkeley: University of California Press, 1997.

Darwin, John. *After Tamerlane: The Rise and Fall of Global Empires, 1400–2000*. London: Penguin, 2008.

de Bruijn, Mirjam, Francis Nyamnjoh, and Inge Brinkman, eds. *Mobile Phones: The New Talking Drums of Everyday Africa*. Bamenda, Cameroon: Langaa RPCIG, 2009.

de Maret, Pierre. "Sanga: New Excavations, More Data and Some Related Problems." *Journal of African History* 18, no. 3 (1977): 321–37.

Denny, S. R. "Val Gielgud and the Slave Traders." *Northern Rhodesia Journal* 3, no. 4 (1957): 331–38.

Deutsch, Jan-Georg. "Notes on the Rise of Slavery and Social Change in Unyamwezi, c. 1860–1900." In *Slavery in the Great Lakes Region of East*

Africa, edited by Henri Médard and Shane Doyle, 76–110. Oxford: James Currey, 2007.

Dibwe dia Mwembu, Donatien. "The Role of Firearms in the Songye Region, 1869–1960." In *The Objects of Life in Central Africa: The History of Consumption and Social Change, 1840–1980*, edited by Robert Ross, Marja Hinfelaar, and Iva Peša, 41–64. Leiden: Brill, 2013.

Dubrunfaut, Paul. "Trade Guns in Africa." In *Fatal Beauty: Traditional Weapons from Central Africa*, 77–85. Taipei: National Museum of History, 2009.

Edgerton, David. "Innovation, Technology, or History? What is the Historiography of Technology About?" *Technology and Culture* 51, no. 3 (2010): 680–97.

Eldredge, Elizabeth A. "Sources of Conflict in Southern Africa, c. 1800–30: The 'Mfecane' Reconsidered." *Journal of African History* 33, no. 1 (1992): 1–35.

Ellis, Stephen. *The Mask of Anarchy: The Destruction of Liberia and the Religious Dimension of an African Civil War*. London: Hurst, 1999.

Ellis, Stephen, and Gerrie ter Haar. *Worlds of Power: Religious Thought and Political Practice in Africa*. New York: Oxford University Press, 2004.

Etherington, Norman. *The Great Treks: The Transformation of Southern Africa, 1815–1854*. Harlow: Longman, 2001.

Fabian, Johannes. *Out of Our Minds: Reason and Madness in the Exploration of Central Africa*. Berkeley: University of California Press, 2000.

Feierman, Steven. "The Shambaa." In *Tanzania before 1900*, edited by Andrew D. Roberts, 1–15. Nairobi: East African Publishing House, 1968.

———. *The Shambaa Kingdom: A History*. Madison: University of Wisconsin Press, 1974.

Findlen, Paula, ed. *Early Modern Things: Objects and Their Histories, 1500–1800*. London: Routledge, 2013.

Finnegan, Ruth H. *Oral Literature in Africa*. 1970. Reprint, Cambridge: Open Book, 2012.

Fisher, Humphrey J., and Virginia Rowland. "Firearms in the Central Sudan." *Journal of African History* 12, no. 2 (1971): 215–39.

Gann, Lewis H. "The End of the Slave Trade in British Central Africa, 1889–1912." *Rhodes-Livingstone Journal* 16 (1954): 27–51.

———. *A History of Northern Rhodesia: Early Days to 1953*. London: Chatto & Windus, 1964.

Gell, Alfred. *Art and Agency: An Anthropological Theory*. Oxford: Oxford University Press, 1998.

Gewald, Jan-Bart, Sabine Luning, and Klaas van Walraven, eds. *The Speed of Change: Motor Vehicles and People in Africa, 1890–2000*. Leiden: Brill, 2009.

Gordon, David M. "The Abolition of the Slave Trade and the Transformation of the South-Central African Interior during the Nineteenth Century." *William and Mary Quarterly* 3rd ser., 66, no. 4 (2009): 915–38.

———. *Invisible Agents: Spirits in a Central African History*. Athens: Ohio University Press, 2012.

————. *Nachituti's Gift: Economy, Society, and Environment in Central Africa.* Madison: University of Wisconsin Press, 2006.

————. "Wearing Cloth, Wielding Guns: Consumption, Trade, and Politics in the South Central African Interior during the Nineteenth Century." In *The Objects of Life in Central Africa: The History of Consumption and Social Change, 1840–1980,* edited by Robert Ross, Marja Hinfelaar, and Iva Peša, 17–39. Leiden: Brill, 2013.

Grant, Jonathan A. *Rulers, Guns, and Money: The Global Arms Trade in the Age of Imperialism.* Cambridge, MA: Harvard University Press, 2007.

Guy, Jeff J. "A Note on Firearms in the Zulu Kingdom with Special Reference to the Anglo-Zulu War, 1879." *Journal of African History* 12, no. 4 (1971): 557–70.

Headrick, Daniel R. *Power over Peoples: Technology, Environments, and Western Imperialism, 1400 to the Present.* Princeton: Princeton University Press, 2010.

————. *The Tools of Empire: Technology and European Imperialism in the Nineteenth Century.* Oxford: Oxford University Press, 1981.

Hecht, Gabrielle. *The Radiance of France: Nuclear Power and National Identity after World War II.* Cambridge, MA: MIT Press, 1998.

Heintze, Beatrix. "Hidden Transfers: Luso-Africans as European Explorers' Experts in Nineteenth-Century West-Central Africa." In *The Power of Doubt: Essays in Honor of David Henige,* edited by Paul S. Landau, 19–40. Madison: Parallel Press, University of Wisconsin Libraries, 2011.

Henige, David. "Truths Yet Unborn? Oral Tradition as a Casualty of Culture Contact." *Journal of African History* 23, no. 3 (1982): 395–412.

Henriques, Isabel de Castro. "Armas de fogo em Angola no século XIX: Uma interpretação." In *Actas de I reunião internacional de história de África: Relação Europa-África no 3° quartel do séc. XIX,* edited by Maria Emilia Madeira Santos, 407–29. Lisbon: Instituto de Investigação Científica Tropical, 1989.

————. *Commerce et changement en Angola au XIXe siècle: Imbangala et Tshokwe face à la modernité.* 2 vols. Paris: L'Harmattan, 1995.

Herbert, Eugenia W. *Red Gold of Africa: Copper in Precolonial History and Culture.* Madison: University of Wisconsin Press, 1984.

Heywood, Linda M. "Slavery and Forced Labor in the Changing Political Economy of Central Angola, 1850–1949." In *The End of Slavery in Africa,* edited by Suzanne Miers and Richard Roberts, 415–36. Madison: University of Wisconsin Press, 1988.

Hicks, Dan. "The Material-Cultural Turn: Event and Effect." In *The Oxford Handbook of Material Culture Studies,* edited by Dan Hicks and Mary C. Beaudry, 25–98. Oxford: Oxford University Press, 2010.

Hogan, Jack. "'Hardly a Place for a Nervous Old Gentleman to Take a Stroll': Firearms and the Zulu during the Anglo-Zulu War." In *A Cultural History of Firearms in the Age of Empire,* edited by Karen Jones, Giacomo Macola, and David Welch, 129–48. Farnham: Ashgate, 2013.

————. "'What Then Happened to Our Eden?': The Long History of Lozi Secessionism, 1890–2013." *Journal of Southern African Studies* 40, no. 5 (2014): 907–24.

Howes, David. "Introduction: Commodities and Cultural Borders." In *Cross-Cultural Consumption: Global Markets, Local Realities*, edited by David Howes, 1–16. London: Routledge, 1996.

Hutchinson, Sharon E. *Nuer Dilemmas: Coping with Money, War, and the State*. Berkeley: University of California Press, 1996.

Iliffe, John. *Honour in African History*. Cambridge: Cambridge University Press, 2005.

———. *A Modern History of Tanganyika*. Cambridge: Cambridge University Press, 1979.

Institut Royal Colonial Belge. *Biographie Coloniale Belge*, vol. 1. Brussels: Librairie Falk Fils, 1948.

Isaacman, Allen F. *Mozambique: The Africanization of a European Institution: The Zambesi Prazos, 1750–1902*. Madison: University of Wisconsin Press, 1972.

Isaacman, Allen F., and Barbara S. Isaacman. *Slavery and Beyond: The Making of Men and Chikunda Ethnic Identities in the Unstable World of South-Central Africa, 1750–1920*. Portsmouth, NH: Heinemann, 2004.

Jourdan, Luca. *Generazione Kalashnikov: Un antropologo dentro la guerra in Congo*. Rome: Laterza, 2010.

———. "Mayi-Mayi: Young Rebels in Kivu, DRC." *Africa Development* 36, nos. 3–4 (2011): 89–111.

Kalusa, Walima T. "Elders, Young Men, and David Livingstone's 'Civilizing Mission': Revisiting the Disintegration of the Kololo Kingdom, 1851–1864." *International Journal of African Historical Studies* 42, no. 1 (2009): 55–80.

Kline, Ronald, and Trevor J. Pinch. "Users as Agents of Technological Change: The Social Construction of the Automobile in the Rural United States." *Technology and Culture* 37, no. 4 (1996): 763–95.

Knight, Ian. *The Anatomy of the Zulu Army: From Shaka to Cetshwayo, 1818–1879*. London: Greenhill, 1995.

Knighton, Ben. "The State as Raider among the Karamojong: 'Where There Are No Guns, They Use the Threat of Guns.'" *Africa: Journal of the International African Institute* 73, no. 3 (2003): 427–55.

Kriger, Colleen E. *Pride of Men: Ironworking in 19th Century West Central Africa*. Portsmouth, NH: Heinemann, 1999.

Laegran, Anne S. "Escape Vehicles? The Internet and the Automobile in a Local-Global Intersection." In *How Users Matter: The Co-Construction of Users and Technology*, edited by Nelly E. J. Oudshoorn and Trevor J. Pinch, 81–100. Cambridge, MA: MIT Press, 2005.

Lamphear, John. "Introduction." In *African Military History*, edited by John Lamphear, xi–xli. Aldershot: Ashgate, 2007.

Langworthy, Harry W. "Introduction: Carl Wiese and Zambezia." In Carl Wiese, *Expedition in East-Central Africa, 1888–1891: A Report*, edited by Harry W. Langworthy, 3–46. Norman: University of Oklahoma Press, 1983.

———. "Swahili Influence in the Area between Lake Malawi and the Luangwa River." *African Historical Studies* 4, no. 3 (1971): 575–602.

Larkin, Brian. *Signal and Noise: Media, Infrastructure, and Urban Culture in Nigeria*. Durham: Duke University Press, 2008.

Leduc-Grimaldi, Mathilde. "'This way!' Aperçu des apports africains aux expéditions européennes du XIXe siècle: Porteurs, éclaireurs et interprètes." In *L'Afrique belge aux XIXe et XXe siècles: Nouvelles recherches et perspectives en histoire coloniale*, edited by Patricia Van Schuylenbergh, Catherine Lanneau, and Pierre-Luc Plasman, 89–99. Brussels: PIE–Peter Lang, 2014.

Legassick, Martin. "Firearms, Horses and Samorian Army Organization 1870–1898." *Journal of African History* 7, no. 1 (1966): 95–115.

Legros, Hugues. *Chasseurs d'ivoire: Une histoire du royaume yeke du Shaba (Zaïre)*. Brussels: Éditions de l'Université de Bruxelles, 1996.

Leslie, Melville E. "Ncheu in the 1890's." *Society of Malawi Journal* 24, no. 1 (1971): 65–78.

Lie, Merete, and Knut H. Sørensen, eds. *Making Technology Our Own? Domesticating Technology into Everyday Life*. Oslo: Scandinavian University Press, 1996.

Linden, Ian. "The Maseko Ngoni at Domwe, 1870–1900." In *The Early History of Malawi*, edited by Bridglal Pachai, 237–51. London: Longman, 1972.

Mackay, Hughie, and Gareth Gillespie. "Extending the Social Shaping of Technology Approach: Ideology and Appropriation." *Social Studies of Science* 22, no. 4 (1992): 685–716.

MacKenzie, John M. *The Empire of Nature: Hunting, Conservation and British Imperialism*. Manchester: Manchester University Press, 1988.

Macola, Giacomo. "Historical and Ethnographical Publications in the Vernaculars of Colonial Zambia: Missionary Contribution to the 'Creation of Tribalism.'" *Journal of Religion in Africa* 33, no. 4 (2003): 343–64.

———. "The History of the Eastern Lunda Royal Capitals to 1900." In *The Urban Experience in Eastern Africa, c. 1750–2000*, edited by Andrew Burton, special issue, *Azania* 14, nos. 36–37 (2002): 31–45.

———. *The Kingdom of Kazembe: History and Politics in North-Eastern Zambia and Katanga to 1950*. Hamburg, Germany: LIT Verlag, 2002.

———. "Reassessing the Significance of Firearms in Central Africa: The Case of North-Western Zambia to the 1920s." *Journal of African History* 51, no. 3 (2010): 301–21.

———. "'They Disdain Firearms': The Relationship between Guns and the Ngoni of Eastern Zambia to the Early Twentieth Century." In *A Cultural History of Firearms in the Age of Empire*, edited by Karen Jones, Giacomo Macola, and David Welch, 101–28. Farnham: Ashgate, 2013.

Mainga, Mutumba. *Bulozi under the Luyana Kings: Political Evolution and State Formation in Pre-Colonial Zambia*. London: Longman, 1973.

Marjomaa, Risto. "The Martial Spirit: Yao Soldiers in British Service in Nyasaland (Malawi), 1895–1939." *Journal of African History* 44, no. 3 (2003): 413–32.

Marks, Shula, and Anthony Atmore. "Firearms in Southern Africa: A Survey." *Journal of African History* 12, no. 4 (1971): 517–30.

Marks, Stuart A. *Large Mammals and a Brave People: Subsistence Hunters in Zambia*. Seattle: University of Washington Press, 1976.

Matthews, T. I. "Portuguese, Chikunda, and Peoples of the Gwembe Valley: The Impact of the 'Lower Zambezi Complex' on Southern Zambia." *Journal of African History* 22, no. 1 (1981): 23–41.

Mavhunga, Clapperton C. "Firearms Diffusion, Exotic and Indigenous Knowledge Systems in the Lowveld Frontier, South Eastern Zimbabwe, 1870–1920." *Comparative Technology Transfer and Society* 1, no. 2 (2003): 201–31.

McCracken, Grant D. *Culture and Consumption: New Approaches to the Symbolic Character of Consumer Goods and Activities*. Bloomington: Indiana University Press, 1988.

McCracken, John. "Coercion and Control in Nyasaland: Aspects of the History of a Colonial Police Force." *Journal of African History* 27, no. 1 (1986): 127–47.

———. *A History of Malawi 1859–1966*. Woodbridge, UK: James Currey, 2012.

———. *Politics and Christianity in Malawi, 1875–1940: The Impact of the Livingstonia Mission in the Northern Province*. Cambridge: Cambridge University Press, 1977.

Miller, Daniel. "Consumption and Commodities." *Annual Review of Anthropology* 24 (1995): 141–61.

———. *Material Culture and Mass Consumption*. Oxford: Blackwell, 1987.

———. "The Young and the Restless in Trinidad: A Case of the Local and the Global in Mass Consumption." In *Consuming Technologies: Media and Information in Domestic Spaces*, edited by Roger Silverstone and Eric Hirsch, 163–82. London: Routledge, 1992.

Miller, Daniel, and Christopher Tilley. "Editorial." *Journal of Material Culture* 1, no. 1 (1996): 5–14.

Miller, Joseph C. "Cokwe Trade and Conquest in the Nineteenth Century." In *Pre-Colonial African Trade: Essays on Trade in Central and Eastern Africa before 1900*, edited by Richard Gray and David Birmingham, 175–201. London: Oxford University Press, 1970.

———. "The Imbangala and the Chronology of Early Central African History." *Journal of African History* 13, no. 4 (1972): 549–74.

———. "Imports at Luanda, Angola: 1785–1823." In *Figuring African Trade: Proceedings of the Symposium on the Quantification and Structure of the Import and Export and Long-Distance Trade of Africa in the Nineteenth Century (c. 1800–1913)*, edited by G. Liesegang, H. Pasch, and A. Jones, 165–246. Berlin: Dietrich Reimer Verlag, 1986.

———. *Kings and Kinsmen: Early Mbundu States in Angola*. Oxford: Clarendon, 1976.

———. *Way of Death: Merchant Capitalism and the Angolan Slave Trade, 1730–1830*. Madison: University of Wisconsin Press, 1988.

Moyd, Michelle R. *Violent Intermediaries: African Soldiers, Conquest, and Everyday Colonialism in German East Africa*. Athens: Ohio University Press, 2014.

Mkutu, Kennedy A. *Guns and Governance in the Rift Valley: Pastoralist Conflict and Small Arms*. Oxford: James Currey, 2008.

Nasson, Bill. "'Give Him a Gun, NOW': Soldiers but Not Quite Soldiers in South Africa's Second World War, 1939–1945." In *A Cultural History of Firearms in the Age of Empire*, edited by Karen Jones, Giacomo Macola, and David Welch, 191–210. Farnham: Ashgate, 2013.

Newitt, Malyn D. D. *Portuguese Settlement on the Zambesi: Exploration, Land Tenure and Colonial Rule in East Africa*. London: Longman, 1973.

Osborne, Myles. "Controlling Development: 'Martial Race' and Empire in Kenya, 1945–59." *Journal of Imperial and Commonwealth History* 42, no. 3 (2014): 464–85.

Oudshoorn, Nelly E. J., and Trevor J. Pinch. "Introduction: How Users and Non-Users Matter." In *How Users Matter: The Co-Construction of Users and Technology*, edited by Nelly E. J. Oudshoorn and Trevor J. Pinch, 1–29. Cambridge, MA: MIT Press, 2005.

Page, Melvin E. "The Manyema Hordes of Tippu Tip: A Case Study in Social Stratification and the Slave Trade in Eastern Africa." *International Journal of African Historical Studies* 7, no. 1 (1974): 69–84.

Papstein, Robert J. "From Ethnic Identity to Tribalism: The Upper Zambezi Region of Zambia, 1830–1981." In *The Creation of Tribalism in Southern Africa*, edited by Leroy Vail, 372–94. London: James Currey, 1989.

Parsons, Timothy H. *The African Rank-and-File: Social Implications of Colonial Military Service in the King's African Rifles, 1902–1964*. Portsmouth, NH: Heinemann, 1999.

———. "'Wakamba Warriors Are Soldiers of the Queen': The Evolution of the Kamba as a Martial Race, 1890–1970." *Ethnohistory* 46, no. 4 (1999): 671–701.

Pélissier, René. *Les guerres grises: Résistance et révoltes en Angola, 1845–1941*. Orgeval, France: Pélissier, 1977.

Pesek, Michael. "*Ruga-ruga*: The History of an African Profession, 1820–1918." In *German Colonialism Revisited: African, Asian, and Oceanic Experiences*, edited by Nina Berman, Klaus Mühlhahn, and Patrice Nganang, 85–100. Ann Arbor: University of Michigan Press, 2014.

Peters, Krijn, and Paul Richards "'Why We Fight': Voices of Youth Combatants in Sierra Leone." *Africa: Journal of the International African Institute* 68, no. 2 (1998): 183–210.

Phillipson, David W. "Gun-Flint Manufacture in North-Western Zambia." *Antiquity* 43, no. 172 (1969): 301–4.

Phimister, Ian. *An Economic and Social History of Zimbabwe, 1890–1948: Capital Accumulation and Class Struggle*. London: Longman, 1988.

Pilossof, Rory. "'Guns Don't Colonise People . . .': The Role and Use of Firearms in Pre-Colonial and Colonial Africa." *Kronos* 36, no. 1 (2010): 266–77.

Pinch, Trevor J., and Wiebe E. Bijker. "The Social Construction of Facts and Artifacts: Or How the Sociology of Science and the Sociology of Technology Might Benefit Each Other." *Social Studies of Science* 14, no. 3 (1984): 399–441.

Poole, E. H. Lane. "Mpeseni and the Exploration Companies, 1885–1898." *Northern Rhodesia Journal* 5, no. 3 (1963): 221–32.

Pratt, Mary Louise. *Imperial Eyes: Travel Writing and Transculturation.* London: Routledge, 1992.

Prestholdt, Jeremy. *Domesticating the World: African Consumerism and the Genealogies of Globalization.* Berkeley: University of California Press, 2008.

Prins, Gwyn. *The Hidden Hippopotamus: Reappraisal in African History: The Early Colonial Experience in Western Zambia.* Cambridge: Cambridge University Press, 1980.

Pritchett, James A. *Friends for Life, Friends for Death: Cohorts and Consciousness among the Lunda-Ndembu.* Charlottesville: University of Virginia Press, 2007.

Ramsay, Jeff. "Firearms in Nineteenth-Century Botswana: The Case of Livingstone's 8-Bore Bullet." *South African Historical Journal* 66, no. 3 (2014): 440–69.

Read, Margaret. *The Ngoni of Nyasaland.* 1956. Reprint, London: Frank Cass, 1970.

———. "Tradition and Prestige among the Ngoni." *Africa: Journal of the International African Institute* 9, no. 4 (1936): 453–84.

Reefe, Thomas Q. *The Rainbow and the Kings: A History of the Luba Empire to 1891.* Berkeley: University of California Press, 1981.

———. "The Societies of the Eastern Savanna." In *History of Central Africa*, vol. 1, edited by David Birmingham and Phyllis M. Martin, 160–204. London: Longman, 1983.

Reid, Richard J. "Past and Presentism: The 'Precolonial' and the Foreshortening of African History." *Journal of African History* 52, no. 2 (2011): 135–55.

———. *Political Power in Pre-Colonial Buganda: Economy, Society and Warfare in the Nineteenth Century.* Oxford: James Currey, 2002.

———. "Revisiting Primitive War: Perceptions of Violence and Race in History." *War and Society* 26, no. 2 (2007): 1–25.

———. "Violence and Its Sources: European Witnesses to the Military Revolution in Nineteenth-Century Eastern Africa." In *The Power of Doubt: Essays in Honor of David Henige*, edited by Paul S. Landau, 41–59. Madison: Parallel Press, University of Wisconsin Libraries, 2011.

———. *Warfare in African History.* Cambridge: Cambridge University Press, 2012.

———. *War in Pre-Colonial Eastern Africa: The Patterns and Meanings of State-Level Conflict in the Nineteenth Century.* Oxford: James Currey, 2007.

Roberts, Allen F. *A Dance of Assassins: Performing Early Colonial Hegemony in the Congo.* Bloomington: Indiana University Press, 2013.

Roberts, Andrew D. "Firearms in North-Eastern Zambia before 1900." *Transafrican Journal of History* 1, no. 2 (1971): 3–21.

———. *A History of the Bemba: Political Growth and Change in North-Eastern Zambia before 1900.* London: Longman, 1973.

———. A History of Zambia. London: Heinemann, 1976.

———. "Livingstone's Value to the Historian of African Societies." In David Livingstone and Africa, edited by Centre of African Studies, 49–67. Edinburgh: University of Edinburgh, Centre of African Studies, 1973.

———. "The Nyamwezi." In Tanzania before 1900, edited by Andrew D. Roberts, 117–50. Nairobi: East African Publishing House, 1968.

Rockel, Stephen J. "'A Nation of Porters': The Nyamwezi and the Labour Market in Nineteenth-Century Tanzania." Journal of African History 41, no. 2 (2000): 173–95.

Roes, Aldwin. "Towards a History of Mass Violence in the Etat Indépendant du Congo, 1885–1908." South African Historical Journal 62, no. 4 (2010): 634–70.

Ross, Andrew C. Blantyre Mission and the Making of Modern Malawi. Blantyre, Malawi: Christian Literature Association in Malawi, 1996.

Sahlins, Marshall. Islands of History. Chicago: University of Chicago Press, 1985.

Santos, Maria Emilia Madeira. "Introdução (trajectória do comércio do Bié)." In Viagens e apontamentos de um portuense em África, edited by Maria Emilia Madeira Santos, 33–216. Coimbra: Biblioteca Geral da Universidade de Coimbra, 1986.

———. "Tecnologias em presença: Manufacturas Europeias e artefactos Africanos (c. 1850–1880)." In Actas de I reunião internacional de história de África: Relação Europa-África no 3° quartel do séc. XIX, edited by Maria Emilia Madeira Santos, 207–36. Lisbon: Instituto de Investigação Científica Tropical, 1989.

Schler, Lynn. "Bridewealth, Guns and Other Status Symbols: Immigration and Consumption in Colonial Duala." Journal of African Cultural Studies 16, no. 2 (2003): 213–34.

Schmidt, Heike. "'Deadly Silence Predominates in this District': The Maji Maji War and Its Aftermath in Ungoni." In Maji Maji: Lifting the Fog of War, edited by James Giblin and Jamie Monson, 183–219. Leiden: Brill, 2010.

Shear, Keith. "'Taken as Boys': The Politics of Black Police Employment and Experience in Early Twentieth-Century South Africa." In Men and Masculinities in Modern Africa, edited by Lisa A. Lindsay and Stephan F. Miescher, 109–27. Portsmouth, NH: Heinemann, 2003.

Sheriff, Abdul. Slaves, Spices and Ivory in Zanzibar: Integration of an East African Commercial Empire into the World Economy, 1770–1873. London: James Currey, 1987.

Spear, Thomas. "Neo-traditionalism and the Limits of Invention in British Colonial Africa," Journal of African History 44, no. 1 (2003): 3–27.

Stahl, Ann B. "Material Histories." In The Oxford Handbook of Material Culture Studies, edited by Dan Hicks and Mary C. Beaudry, 148–70. Oxford: Oxford University Press, 2010.

Stapleton, Timothy J. African Police and Soldiers in Colonial Zimbabwe, 1923–80. Rochester, NY: University of Rochester Press, 2011.

Steinhart, Edward I. *Black Poachers, White Hunters: A Social History of Hunting in Colonial Kenya.* Oxford: James Currey, 2006.

Stengers, Jean, and Jan Vansina. "King Leopold's Congo, 1886–1908." In *The Cambridge History of Africa*, vol. 6, c. 1870–c. 1905, edited by Roland Oliver and G. N. Sanderson, 315–58. Cambridge: Cambridge University Press, 1985.

Storey, William K. *Guns, Race, and Power in Colonial South Africa.* Cambridge: Cambridge University Press, 2008.

Streets, Heather. *Martial Races: The Military, Race and Masculinity in British Imperial Culture, 1857–1914.* Manchester: Manchester University Press, 2004.

Thomas, Nicholas. *Entangled Objects: Exchange, Material Culture, and Colonialism in the Pacific.* Cambridge, MA: Harvard University Press, 1991.

Thompson, T. Jack. "The Origins, Migration, and Settlement of the Northern Ngoni." *Society of Malawi Journal* 34, no. 1 (1981): 6–35.

Thornton, John K. "The Art of War in Angola, 1575–1680." *Comparative Studies in Society and History* 30, no. 2 (1998): 360–78.

———. *Warfare in Atlantic Africa, 1500–1800.* London: UCL Press, 1999.

Tylden, G. "The Gun Trade in Central and Southern Africa." *Northern Rhodesia Journal* 2, no. 1 (1953): 43–48.

Vail, Leroy, ed. *The Creation of Tribalism in Southern Africa.* London: James Currey, 1989.

———. "The Making of the 'Dead North': A Study of the Ngoni Rule in Northern Malawi, c. 1855–1907." In *Before and After Shaka: Papers in Nguni History*, edited by Jeffrey B. Peires, 230–67. Grahamstown, South Africa: Institute of Social and Economic Research, Rhodes University, 1981.

Van Onselen, Charles. "The Role of Collaborators in the Rhodesian Mining Industry, 1900–1935." *African Affairs* 72, no. 289 (1973): 401–18.

Vansina, Jan. *Being Colonized: The Kuba Experience in Rural Congo, 1880–1960.* Madison: University of Wisconsin Press, 2010.

———. *The Children of Woot: A History of the Kuba Peoples.* Madison: University of Wisconsin Press, 1978.

———. "For Oral Tradition (But Not against Braudel)." *History in Africa* 5 (1978): 351–56.

———. *How Societies Are Born: Governance in West Central Africa before 1600.* Charlottesville: University of Virginia Press, 2004.

———. "It Never Happened: Kinguri's Exodus and Its Consequences." *History in Africa* 25 (1998): 387–403.

———. "Long-Distance Trade-Routes in Central Africa." *Journal of African History* 3, no. 3 (1962): 375–90.

Vellut, Jean-Luc. "L'économie internationale des côtes de Guinée Inférieure au XIXe siècle." In *Actas de I reunião internacional de história de África: Relação Europa-África no 3° quartel do séc. XIX*, edited by Maria Emilia Madeira Santos, 135–201. Lisbon: Instituto de Investigação Científica Tropical, 1989.

———. "Garenganze/Katanga–Bié–Benguela and Beyond: The Cycle of Rubber and Slaves at the Turn of the 20th Century." *Portuguese Studies Review* 19, nos. 1–2 (2011): 133–52.

———. "Notes sur le Lunda et la frontière luso-africaine (1700–1900)." *Études d'Histoire Africaine* 3 (1972): 61–166.

———. "La violence armée dans l'État Indépendant du Congo: Ténèbres et clartés dans l'histoire d'un état conquérant." *Cultures et Développement* 16, nos. 3–4 (1984): 671–707.

Velsen, Jaap van. "Notes on the History of the Lakeside Tonga of Nyasaland." *African Studies* 18, no. 3 (1959): 105–17.

Vlassenroot, Koen. "A Societal View on Violence and War: Conflict and Militia Formation in Eastern Congo." In *Violence, Political Culture and Development in Africa*, edited by Preben Kaarsholm, 49–65. Oxford: James Currey, 2006.

von Oppen, Achim. *Terms of Trade and Terms of Trust: The History and Contexts of Pre-Colonial Market Production around the Upper Zambezi and Kasai*. Munich: LIT Verlag, n.d. [1994?].

White, Charles M. N. *The Material Culture of the Lunda-Lovale Peoples*. Occasional Papers of the Rhodes-Livingstone Museum 3. Livingstone, Zambia: Rhodes-Livingstone Museum, 1948.

White, Landeg. *Magomero: Portrait of an African Village*. Cambridge: Cambridge University Press, 1987.

Willis, Justin. *Potent Brews: A Social History of Alcohol in East Africa, 1850–1999*. Oxford: James Currey, 2002.

Wilson, K. B. "Cults of Violence and Counter-Violence in Mozambique." *Journal of Southern African Studies* 18, no. 3 (1992): 527–82.

Wright, John. "Turbulent Times: Political Transformations in the North and East, 1760s–1830s." In *The Cambridge History of South Africa*, vol. 1, *From Early Times to 1885*, edited by Carolyn Hamilton, Bernard K. Mbenga, and Robert Ross, 211–52. Cambridge: Cambridge University Press, 2010.

Wright, Marcia, and Peter Lary. "Swahili Settlements in Northern Zambia and Malawi." *African Historical Studies* 4, no. 3 (1971): 547–73.

Wright, Tim B. *The History of the Northern Rhodesia Police*. Bristol: British Empire & Commonwealth Museum Press, 2001.

Youngs, Tim. *Travellers in Africa: British Travelogues, 1850–1900*. Manchester: Manchester University Press, 1994.

Unpublished Dissertations and Papers

Hermitte, Eugene L. "An Economic History of Barotseland, 1800–1940." PhD diss., Northwestern University, 1974.

Hogan, Jack. "The Ends of Slavery in Barotseland, Western Zambia (c. 1800–1925)." PhD diss., University of Kent, 2014.

Hoover, J. Jeffrey. "The Seduction of Ruwej: Reconstructing Ruund History (The Nuclear Lunda: Zaire, Angola, Zambia)." 2 vols. PhD diss., Yale University, 1978.

Kalenga, Pierre. "Situation socio-politique du Katanga ancien (1870–1911)." Mémoire de maîtrise, University of Lubumbashi, DR Congo, 2010.

Katompa, K. Asani bin. "L'opposition Sanga à Msiri et à l'administration coloniale Belge (1891–1911)." Mémoire de licence en histoire, Université Nationale du Zaïre, Lubumbashi, 1977.

Kennes, Erik. "Fin du cycle post-colonial au Katanga, RD Congo: Rébellions, sécession et leurs mémoires dans la dynamique des articulations entre l'État central et l'autonomie régionale, 1960–2007." PhD diss., Université Laval and Université Paris I, 2009.

Mavhunga, Clapperton C. "The Mobile Workshop: Mobility, Technology, and Human-Animal Interaction in Gonarezhou (National Park), 1850–present." PhD diss., University of Michigan, 2008.

Papstein, Robert J. "The Upper Zambezi: A History of the Luvale People, 1000–1900." PhD diss., UCLA, 1978.

Phiri, Kings M. "Chewa History in Central Malawi and the Use of Oral Tradition, 1600–1920." PhD diss., University of Wisconsin–Madison, 1975.

Rau, William E. "Mpezeni's Ngoni of Eastern Zambia, 1870–1920." PhD diss., UCLA, 1974.

Rennie, John K. "The Conquest of Maala in 1901: An Exercise in Oral and Documentary Evidence among the Ila of Namwala District, Zambia." Paper presented to the Institute for African Studies and the Department of History, University of Zambia, Lusaka, February 1982.

Williams-Myers, Albert J. "The Nsenga of Central Africa: Political and Economic Aspects of Clan History, 1700 to the Late Nineteenth Century." PhD diss., UCLA, 1978.

Index

as "the coward's weapon," 136 (*see also* Ngoni people: close combat and)

elephant, 59, 66, 68

"gun society," 20, 24, 57, 73, 87, 94, 95, 114, 163; defined, 19

Kalashnikov (AK-47) rifles, 164

Martini-Henry rifles, 65, 69, 97, 157, 160

Maxim, 143, 146, 148

"medicine," 62

muskets, 80, 82, 90, 127; as accessible technology, 61, 73, 106; ammunition for, 61, 67, 80, 87, 96, 98, 100, 101, 102, 103, 123, 130, 133; barrels of, 55, 61, 64, 66, 87, 113; Belgian-made, 43, 54–55, 106; central African terms for, 24, 61, 70, 71–72, 78; chronology of arrival of, 12, 20, 30, 39–40, 41, 43, 44, 53, 58–60, 63, 66, 70–71, 77–81, 86; as currency, 15, 53, 72, 73, 100, 103–4, 105; environmental impact of, 54, 62, 66, 84, 100, 101; in female hands, 72, 103 (*see also* masculinity); figures pertaining to, 40, 66, 69, 81, 86, 98, 101, 103, 112, 131, 132–33, 165; flintlock, 40, 54, 61, 66, 69, 75, 78, 79, 81, 87, 93, 97, 98, 109, 112, 123, 128, 162, 165; homemade, 1–4, 48, 112–14, 115; *lazarinas*, 54–55, 60–64, 65, 70, 81, 85, 98, 106, 114; as means of political centralization, 29, 46, 53, 69–70, 73, 74, 83, 88–89, 93, 95–96, 97, 162; military uses of, 20, 40, 44–45, 46, 49, 62–63, 64, 66, 68, 69, 71, 73, 74–75, 78, 79, 83, 84–87, 89, 92, 93, 95, 101, 111, 115, 123-25, 130, 131, 150, 162 (*see also* war, warfare); percussion-lock, 1, 66, 75, 83, 85, 86, 93, 97, 109, 110, 112, 113, 114, 123, 165; Portuguese-made, 54, 63; repair, 15, 56, 58–59, 61, 67–68, 87; symbolic attributes of, 3, 15, 20, 21, 40, 43, 53, 57, 64–65, 70, 73, 114, 132, 162; technical limitations of, 11–12, 20, 43, 54, 56–57, 62, 67–68, 73, 119, 123, 133; as tools of production, 15, 43, 49, 53, 62, 73, 74–75, 83–84, 88–89, 93, 95–96, 100, 103, 114, 162 (*see also* hunting); trade, 40, 43, 54–55, 56, 129

rejection of. *See* technological disengagement

scholarship on, 11–13, 54, 161–62, 166–67

Snider-Enfield rifles, 127, 129, 151

spiritual appraisals of, 15

See also marksmanship; masculinity; trade: in guns

Hall, P. E., 103–4

Harding, Colin, 65, 153–55

Hawes, Albert George Sidney, 129

Hazell, William, 102

Headrick, Daniel R., 8

Hecht, Gabrielle, 5

Hole, H. Marshall, 158

Holub, Emil, 66, 67, 68–69

Hoover, J. Jeffrey, 33

Howes, David, 7

hunting, 3, 11, 15, 38, 48, 49, 53, 60–62, 64, 67, 69, 71–72, 95, 100, 101, 103, 104, 106, 111–14, 133, 162; elephant, 54, 59, 62, 66, 67, 83–84, 100, 107, 111, 112, 129, 156 (*see also* Chikunda people; Chokwe people; trade: ivory); European, 13, 16; game laws, 95, 96, 106, 111, 114, 163; rituals and ceremonies, 15, 66, 72; trapping and, 114; warrior ethos and, 137–38. *See also* guns: muskets: as tools of production

ibn Habib, Said, 40, 80

ibn Saleh, Muhammed, 80

Ila people, 38, 40, 44, 68, 69, 70, 71, 100, 119

Iliffe, John, 136

Imbangala people, 34, 44

imigubo. See songs

iron, ironworking, 31, 32, 56, 61, 66, 67–68, 73, 87, 101. *See also* guns: muskets: repair

Isaacman, Allen F. and Barbara S., 48–49, 130

Ivens, Roberto, 81–82, 83–84, 85, 89, 111

ivory. *See* trade: ivory

Jenniges, Jean-Mathieu, 110

Johnston, Harry H., 152

Kabila, Laurent D., 1

Kabobo, 83

Kakenge, 38, 60, 65, 98

Kalala, 108

Kalenga, Pierre, 1

Kalunkumia. See under *Mpande*

Kalusa, Walima T., 58

Manning, William H., 146
Marenga, 124–25
marksmanship, 22, 62, 157–58; reportedly poor, 58, 63
Masangu, 67
masculinity: guns and, 3, 12, 15, 43, 48, 53, 65, 70, 73, 98, 114, 160, 162, 164. *See also* Ngoni people: honor and masculinity among
Masengo, 89
Mashonaland Native Police (MNP), 153, 154, 156
Mashukulumbwe people. See Ila people
Matakenya, 49
Maugham, Reginald C. F., 150–51
Mavhunga, Clapperton C., 13, 57
Mburuma, 130
Mbwela people, 61
McGregor, G. A., 99
Melland, Frank H., 71
Mfecane, 49–50, 58, 59, 120, 162
Milambo Myelemyele, 87
Miller, Joseph C., 15, 54, 56
missionaries, 10, 23, 37, 38, 65, 68, 69, 72, 105, 111, 130, 149. *See also* Free Church of Scotland; Plymouth Brethren
Mkanda, 121, 131
Mkutu, Kennedy A., 166
Mlonyeni, 147
M'Mbelwa I, 120–22, 123, 124, 126–27, 128, 134–35, 141
Mobutu Sese Seko, 1
Moloney, Joseph A., 78
Mozambique, 18, 41, 48, 49, 59, 121, 130, 134, 167
Mpande, 76; Kalunkumia, 89, 91; Kya Mulemba, 77, 89
Mpepe, 58
Mperembe, 124, 135
Mpezeni I, 120–22, 124, 130–33, 134, 136, 137, 141, 142–46, 147, 149, 150, 151, 152, 153, 155
Mponda, 128; II Nkwate, 128, 129–30
Mputa, 122
Msiri, 1, 3, 15, 45, 46, 49, 71, 74–93 passim, 95, 106, 107, 108, 110, 111, 112, 115, 161, 162
Mthetwa people, 50
Mtwalo, 126
Mufunga, 93
Mukanda Bantu, 77–78, 80, 89, 91, 92, 93, 107–8, 109–11, 115
Mukemwa, 108

Mulowanyama, 83, 90, 92, 93, 109
Munongo, Antoine Mwenda, 79, 87
Munongo, Godefroid, 114
Mushima, 100; Mubambe, 71
muskets. *See under* guns
Mutwila, 91, 108
Mwant Yavs, 32–34, 44; Mudib, 44; Mukaz, 44; Muteb, 44
Mwase Kasungu, 123, 124, 131, 136, 143
Mwashya, 83, 91
Mwata Kazembes, 39, 41; IV Keleka, 36, 40; VI Chinyanta Munona, 78; VII Muonga Sunkutu, 77, 80; IX Lukwesa Mpanga, 86; X Kanyembo Ntemena, 86
Mzilikazi, 59

Ndebele people, 59, 69, 101, 152
Ndwandwe people, 50, 122
Ngandu, 86
Ngoni people, 19, 50, 59, 162–63; age-set regiments among, 15, 20, 50, 122, 133, 134, 138, 144, 145, 147; captives among, 122, 124, 126, 128, 133–34, 138, 139–40; close combat and, 15, 20, 21, 120, 129, 136, 142, 147, 150, 157; early history of, 120–22; European subjugation of, 21, 141–51
honor and masculinity among, 15, 20, 21, 120, 135–37, 138–39, 142, 144, 147, 151–52, 153–54, 157, 160, 162; and long-distance trade, 50, 119–20, 126–28, 132, 162; as "martial race," 21, 141–42, 152–53, 156–57; military setbacks by, 123–25, 129–31, 133. *See also* Barotse Native Police (BNP): Ngoni in; technological disengagement
Ng'onomo, 124, 125, 131, 134
Nguni: language, 122; people, 49–50, 133
Ngwane people, 50
Nigeria, 8, 9
North Charterland Exploration Company (NCEC), 133, 143–44, 146, 151, 152
North-Eastern Rhodesia. *See under* Rhodesia
North-Eastern Rhodesia Constabulary, 154, 155
Northern Rhodesia. *See under* Rhodesia
North-Western Rhodesia. *See under* Rhodesia
Nsenga people, 49, 121, 130, 132
Nsingo, 144–49